MECHANICAL TRADES
POCKET MANUAL

by Carl A. Nelson

THEODORE AUDEL & CO.

a division of

HOWARD W. SAMS & CO., INC.

4300 West 62nd Street
Indianapolis, Indiana 46268

FIRST EDITION

Seventh Printing—1981

International Standard Book Number: 0-672-23215-4
Library of Congress Catalog Card Number: 74-91639

Foreword

The intention of this manual is to provide reference material for mechanical tradesmen. While it is primarily concerned with installation, maintenance and repair of machinery and equipment, other fields of activity involved in this overall operation are included. This broad range of information on methods, procedures, equipment, tools etc., is presented in convenient form and plain language to aid the mechanic in performance of day to day tasks.

The constant aim has been to present the subject as clearly, concisely and simply as possible. To accomplish this, numerous sketches and practical examples are used, and the explanations are given as briefly and simply as possible. Discussions of background principles and theory have been limited to the essentials required for understanding of the subject matter.

Information is given on a wide range of subjects, including many new materials and methods. While intended primarily as an industrial mechanic's handbook, the data, information, directions, etc., apply to operations performed by many mechanical trades.

CARL A. NELSON

Contents

	Pages
Mechanical Drawing	9-13
Isometric Drawing	14
Single Line Isometric Pipe Drawing	15-19
Field Layout	20-23
Arc layout method—3–4–5 layout method	
Machinery Installation	24-26
Leveling—determining shim thickness—elevation	
Machine Assembly	27-30
Allowance for fits—keys, key seats—keyways—scraping—flat surface scraping	
Torque Wrench	31
Units of measure	
Torque Wrench Terms	32-33
Push and pull—break-away torque—set or seizure—run down resistance—wrench size—torque and tension—attachments	
Torque Specifications	34
Torque values for steel fasteners	

Mechanical Power Transmission 35-38

    Pitch—pitch diameter and pitch circle—calculation—calculation examples

"V" Belts 39-50

    Fractional horsepower—standard multiple—wedge—"V" belt matching—"V" belt drive alignment—"V" belt alignment—"V" belt installation—checking belt tension—wedge belt tension—Browning belt tension checker—"V" belt replacement—sheave groove wear—sheave groove

Chain Drives 51-59

    Roller chain—roller chain dimensions—standard roller chain numbers—roller chain connections—roller chain sprockets—roller chain installation—silent chain—chain replacement

Spur Gears 60-66

    Pitch diameter and center distance—circular pitch—diametral pitch—diametral pitch approximation—gear tooth parts—gear dimensions—backlash

Couplings 67-73

    Coupling alignment—vertical face alignment correction—selection of unit to be adjusted—the indicator method—temperature change compensation

Screw Threads 74-81

    Coarse–thread series—fine–thread series—extra–fine thread series—constant–pitch series—thread classes—unified thread designation—screw thread terms—translation threads—square thread—acme thread—buttress thread—metric threads—thread tapping

Mechanical Fasteners 82-86

    Coarse threads—fine threads—washers—nuts and pins—characteristics—measurements—retaining rings

Packings and Seals 87-105

    Stuffing box packings—stuffing box arrangement—packing installation

—stuffing box lantern rings—mechanical seals—mechanical–seal types—mechanical seal installation—installation precautions—"O" rings—"O" ring dash number system—formed and molded packings—packing materials—packing installation

Bearings 106-122

Bearing nomenclature—plain bearings—lubrication—bearing failures—antifriction bearings—self-aligning bearings—ball bearing dimensions—ball bearing numbering systems—cylindrical roller bearings—standard ball bearing sizes—spherical roller bearings—mounting procedure—taper roller bearings—fixed and floating bearings—basic adjusting devices

Pumps 123-125

Centrifugal pumps—reciprocating pumps—rotary pumps

Structural Steel 126-138

American standard angles—American standard beams—American standard channels—American standard wide flange beams and columns—simple square–framed beams—clearance cuts—square–framed connections—angle connecting legs—fabrication terms—connection hole locations—dimensions of two–angle connections—steel elevation

Twist Drills 139-141

Drill terms—drill sharpening—freehand drill point grinding

Stair Layout 142-147

Conditions—dimensions—examples—stair length—stair horse layout—stair stringer layout—stair drop

Rigging 148-160

Weight estimating—wire rope—factor of safety—wire rope attachment—multiple reeving—how to measure wire rope—wire rope slings—eyebolts and shackles—knots

Piping 161-172

Thread designations—American standard taper pipe threads—pipe mea-

surement—copper water tube—45-degree offset—45-degree rolling offset—flanged pipe connections—pipe-flange bolt—hole layout—valves

Steam Traps 173-179
   Steam trap testing

Automatic Sprinkler Systems 180-186
   Opening steam supply valve—wet sprinkler systems—operation of wet
   sprinkler alarm system—dry-pipe sprinkler systems—operation of dry-
   pipe sprinkler systems

Carpentry 187-200
   Commercial lumber sizes—building layout—wood-frame building
   construction

Electricity 201-210
   Electrical terms—electrical calculations—electrical circuits—electrical
   wiring

Shop Geometry 211-216
   Lines—angles—circles—triangles—quadrilaterals—regular polygons—
   geometrical construction—dividing a line—erecting a perpendicular
   line—polygon construction—dimensions of polygons and circles

Shop Trigonometry 217-219
   Right-angle triangles—sum of squares

Appendix 220-256

MECHANICAL DRAWING

Mechanical drawing is a graphic language used to convey information, size, location, accuracy, etc., of mechanisms ranging from the simple to the most complex. Because there is a general standardization of techniques throughout the world, this language is understood in spite of native tongue and measurement differences.

To understand this graphic mechanical language requires an understanding of basic principles and concepts as well as knowledge of the conventions commonly followed. Lines define the shape, size, and details of an object, and through the correct use of lines it is possible to graphically describe an object so that it can be accurately visualized. A listing of these lines is called an "alphabet" of lines (Fig. 1.).

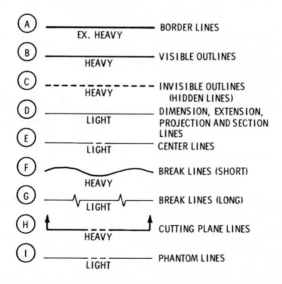

Fig. 1.

The standard system of mechanical drawing is called *Orthographic Projection*. This simply means that the lines of sight are projected parallel and at right angles to the surface viewed, as shown in Fig. 2.

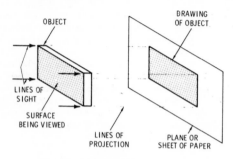

Fig. 2.

To show size shape and location a number of "views" are necessary. Each view represents the true shape and size of the *surface* of the object being viewed, as it is seen by looking directly at it. They are arranged so that each view represents the surface adjacent to it. The top view is drawn above and vertically in line with the front view. The side views, at the side of the front view and horizontally in line, to the right or left as the case may be. The three view drawing (usually showing the top, front, and right side) is the most widely used combination of views. The six principal views used in orthographic projection are shown in Fig. 3.

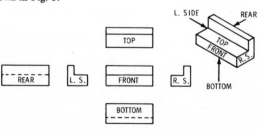

Fig. 3.

As the internal details of a part become more complex and the hidden lines become more numerous, a point is reached where the drawing is difficult to interpret. A technique to simplify such a drawing is to cut away a portion and expose the inside surfaces. A view with this imaginary cut to reveal inside portions is called a sectional view. Such a view may be a *full section* where the imaginary cutting plane passes completely through the object, or a part section in which the plane extends only part way through the object.

On cutaway section views the invisible edges become visible and may be represented by solid-object lines. The exposed surfaces through which the imaginary cut is made are identified by slant lines called *section* or *cross-hatch* lines. To indicate the edge of the cutting plane and the directions in which the section is viewed, a cutting-plane line is used. Letters are usually placed near the direction arrowheads to identify the section (Fig. 4).

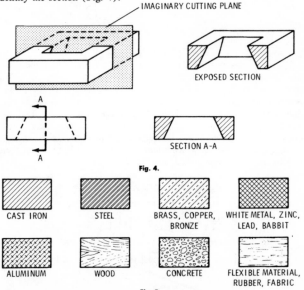

IMAGINARY CUTTING PLANE

EXPOSED SECTION

A

A

SECTION A-A

Fig. 4.

CAST IRON

STEEL

BRASS, COPPER, BRONZE

WHITE METAL, ZINC, LEAD, BABBIT

ALUMINUM

WOOD

CONCRETE

FLEXIBLE MATERIAL, RUBBER, FABRIC

Fig. 5.

Screw threads in one form or another are used on practically all kinds of mechanical objects. They are most widely used on fasteners, adjusting devices and to transmit power. For these purposes a number of thread forms are used, the most common of which are illustrated in Fig. 6.

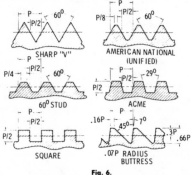

Fig. 6.

Because screw threads are so widely used there is a repeated necessity to show and specify them on mechanical drawings. As a true representation of a screw thread is an extremely laborious operation, it is almost never done. Instead, threads are given a symbolic representation suitable for general understanding. Three methods are used, *Detailed Representation,* which approximates the true appearance, *Schematic Representation,* which is nearly as pictorial and much easier to draw, and *Simplified Representation,* which is the easiest to draw therefore the most commonly used. See Fig. 7.

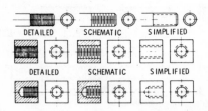

Fig. 7.

Drawings made up of lines to describe shape and contour, must also have dimensions and notes to supply sizes and location. Dimensions are placed between a combination of points and lines. The dimension line indicates the direction in which the dimension applies, and the extension lines refer the dimension to the view. Leaders are used with notes to indicate the feature on the drawing to which the note applies.

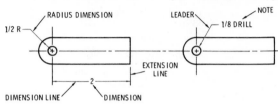

Fig. 8.

Three systems of writing dimension values are in general use. Most widely used is the "common-fraction" system, with all dimension values written as units and common fractions. The second system uses decimal fractions when distances require precision greater than plus or minus ⅟₆₄ of an inch. The third system is the "complete decimal system" and uses decimal fractions for all dimensional values. Two place decimals are used where common fractions are used in the other two systems. When greater precision is required the value is written in three, four, or more places as required. The complete decimal system is increasing in use particularly for machine parts, tools and other precise mechanical type drawings.

When large values must be shown with the common fraction system, feet and inch units may be used. The foot (′) and inch (″) marks may be used to identify the units. Use of feet and inch dimensions should only be used for distances exceeding 72 inches. When dimensions are all in inches, the inch marks are preferably omitted from all dimensions and notes.

The *Unified Standard* thread, which incorporates the earlier National, American, SAE, and ASME standards is the most widely used thread form. It is designated on drawings by a note specifying in sequence the nominal size, number of threads per inch, thread series symbol and the thread class number. The symbols used are (UNC) denoting a coarse thread series, (UNF) denoting a fine thread series, and (UNEF) denoting an extra fine thread series,. Thread class number 1 indicates a loose free fit, number 2 indicates a free fit with very little looseness, number 3 very close fit. The number 2 fit is used for most general applications. The letter A indicates an external thread and the letter B an internal thread (Fig. 9).

NOMINAL SIZE — THREAD SERIES

1/4-20 UNC-2A

THREADS PER INCH — CLASS FIT

Fig. 9.

ISOMETRIC DRAWING

The isometric system of drawing is three-dimensional and more pictorial than the orthographic system. The word isometric means equal measure and has reference to the isometric position which is the basis of isometric drawing.

The isometric position may be developed by rotating a cube around its vertical axis and tilting it forward until all faces are foreshortened equally as shown in Fig. 10A. The overall outline is then a regular hexagon as shown in Fig. 10B. The three lines of the front corner of the cube in isometric position make equal angles with each other and are called the *isometric axes,* as shown in Fig. 10C.

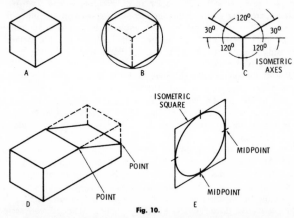

Fig. 10.

On isometric drawings, horizontal object lines are drawn parallel to the 30-degree isometric axis and the vertical lines are drawn parallel to the vertical axis. Actual lengths are shown to scale, frequently resulting in a distorted appearance due to the foreshortening effect. Lines that are not parallel to the isometric axes do not appear in their true length. To draw such lines their ends are located on isometric lines and the points connected as shown in Fig. 10D. A circle shows in isometric as an ellipse, and may be constructed by first making an isometric square, as shown in Fig. 10E. The midpoint of each side is located and the arcs of circles drawn to be tangent at the midpoints, as shown in Fig. 10E.

When making sketches of objects having nonisometric surfaces it may be helpful to imagine the object as contained in rectangular boxes. By sketching the rectangular boxes in isometric position points may be accurately located and construction simplified (Fig. 11).

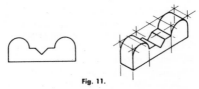

Fig. 11.

SINGLE LINE ISOMETRIC PIPE DRAWING

The rectangular box system is also employed in making isometric pipe drawings. The piping systems are considered as being on the surface or contained inside the box. To simplify the system, pipe is represented by a single line and fittings are shown by symbols (Fig. 12).

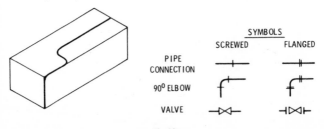

Fig. 12.

To aid in clarifying the drawing and to avoid confusion of direction when making it, an orientation diagram is used. This diagram is a representation of the isometric axes each labeled to indicate relative position and direction. The vertical axis is always labeled "UP" and "DOWN." The two horizontal axes are given appropriate direction labels such as "FRONT," "BACK," "NORTH," "SOUTH," etc.

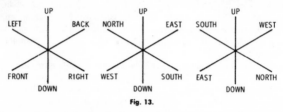

Fig. 13.

As was the case in developing the isometric position, the point of observation for an isometric pipe drawing is always directly in front of the object. In respect to the orientation diagram the observation point is directly in front of the vertical "up" and "down" axis. When the point has been selected from which the pipe system will be viewed, the diagram is labeled to conform to this point. The lines on the drawing are then made to indicate the actual direction of the pipe.

It is important when selecting a viewing position to consider which position will result in a drawing with the most clarity. Two drawings of the same system when viewed from different points will result in one being much easier to read than the other as illustrated in Fig. 14.

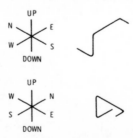

Fig. 14.

Chart 1. Pipe Fitting Symbols

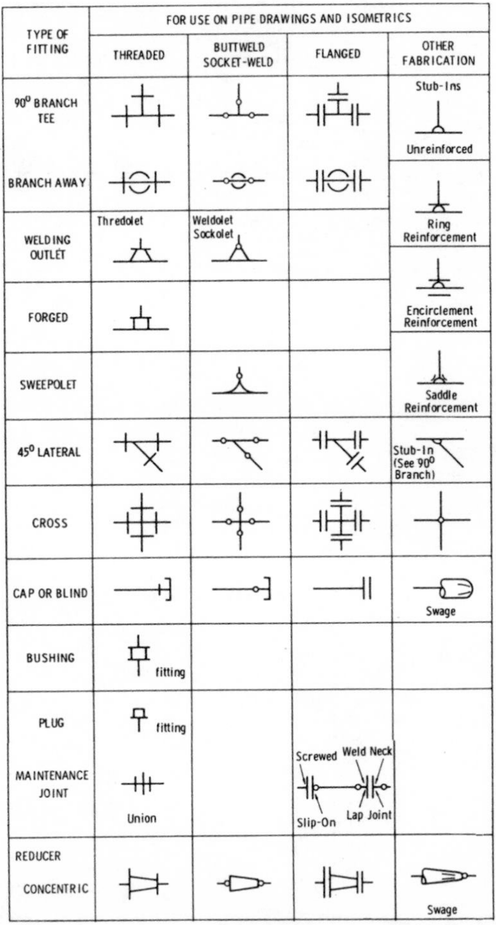

TYPE OF FITTING	FOR USE ON PIPE DRAWINGS AND ISOMETRICS			
	THREADED	BUTTWELD SOCKET-WELD	FLANGED	OTHER FABRICATION
90° BRANCH TEE				Stub-Ins
BRANCH AWAY				Unreinforced
WELDING OUTLET	Thredolet	Weldolet Sockolet		Ring Reinforcement
FORGED				Encirclement Reinforcement
SWEEPOLET				Saddle Reinforcement
45° LATERAL				Stub-In (See 90° Branch)
CROSS				
CAP OR BLIND				Swage
BUSHING	fitting			
PLUG	fitting			
MAINTENANCE JOINT	Union		Screwed Weld Neck Slip-On Lap Joint	
REDUCER CONCENTRIC				Swage

Chart 1. Pipe Fitting Symbols (Continued)

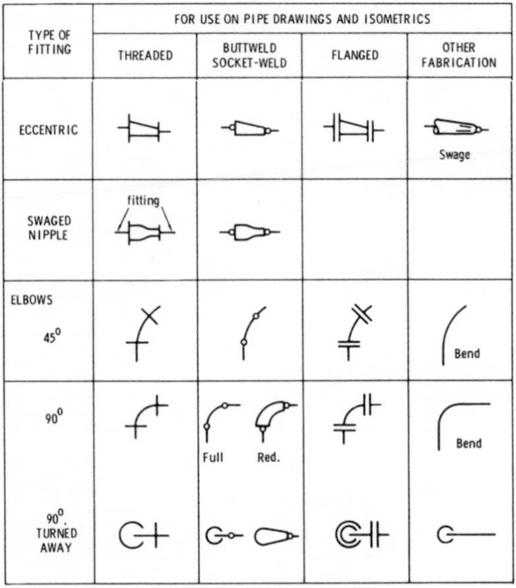

TYPE OF FITTING	FOR USE ON PIPE DRAWINGS AND ISOMETRICS			
	THREADED	BUTTWELD SOCKET-WELD	FLANGED	OTHER FABRICATION
ECCENTRIC				Swage
SWAGED NIPPLE	fitting			
ELBOWS 45°				Bend
90°		Full Red.		Bend
90°, TURNED AWAY				

Chart 2. Valve Symbols

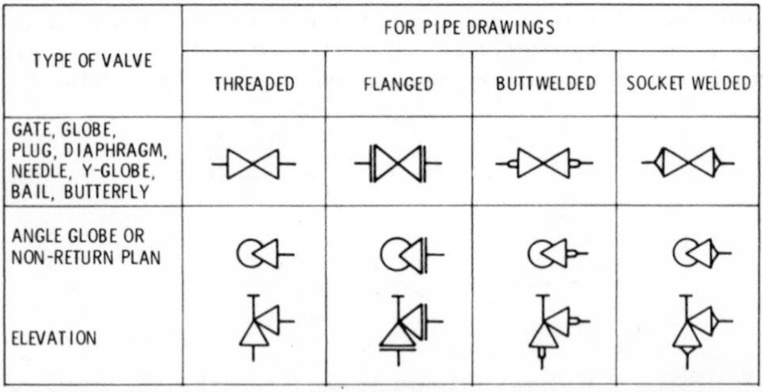

TYPE OF VALVE	FOR PIPE DRAWINGS			
	THREADED	FLANGED	BUTTWELDED	SOCKET WELDED
GATE, GLOBE, PLUG, DIAPHRAGM, NEEDLE, Y-GLOBE, BAIL, BUTTERFLY				
ANGLE GLOBE OR NON-RETURN PLAN				
ELEVATION				

Chart 2. Valve Symbols (Continued)

TYPE OF VALVE	FOR PIPE DRAWINGS			
	THREADED	FLANGED	BUTTWELDED	SOCKET WELDED
SWING, LIFT, TILT OR WAFER TYPE CHECK				
THREE-WAY PLUG VALVE				

The direction of the crossing lines or symbol marks depends on the direction of the pipe. Flange faces on horizontal pipe runs are vertical, therefore fitting marks on horizontal runs should be "Up" and "Down." Vertical pipe run flange faces are horizontal and may be drawn on either horizontal axis.

POOR PRACTICE   GOOD PRACTICE

Fig. 15.

Isometric extension and dimension lines are drawn parallel to the isometric axis. Dimensions may extend to object lines but preferred practice is to have arrowheads end on extension lines. If possible all extension and dimension lines should be placed outside the object.

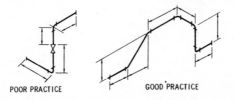

POOR PRACTICE   GOOD PRACTICE

Fig. 16.

FIELD LAYOUT

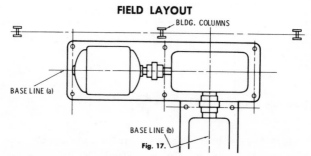

Fig. 17.

The installation of machinery and equipment usually requires its location in respect to given points, objects, or surfaces. These may be building columns, walls, machinery, equipment, etc. To accomplish this in field layout work *base lines* are located in respect to these references and the layout developed from the base lines. For plane surface layout, two base lines at right angles are usually sufficient. Specific lines and points are then located by laying out *center lines* parallel to these base lines. A common field layout problem therefore is layout of right angle base lines.

In the example above, the base line (a) is laid out by measurement parallel to the building columns, and base line (b) is laid out at right angles to base line (a). The center lines for machinery installation are then laid out parallel to the base lines.

The large steel square, or carpenter framing square, while suitable for relatively small layout work should not be used for large scale work. A slight error at the square may be magnified to unacceptable proportions when lines are extended. A more dependable method and accurate method is to develop the layout to suit the size of the job.

There are several right-angle layout methods that will give accurate results when used for large-scale field layout. The two most suitable methods are development by swinging arcs and the 3-4-5 triangle layout. In either case only simple measurements made with care are required. However, the larger the scale of the layout the greater the accuracy since the effect of small errors diminishes as dimensions are increased.

Arc Layout Method

Use a steel tape, wire, or similar device that cannot stretch. Do not use a string line. Reasonable care in executing the following steps are necessary to obtain accurate results. See Fig. 18.

Step #1—Locate two points, (a) and (b), on the straight line at equal distances from the given point. This may be done by measurement, using a marked stick, or swinging a steel tape.

Step #2—Swing arcs from points (a) and (b) using a radius length about 1½ times the measurement used to locate these points. Exactly the same length radius must be used for both arcs. Locate point (c) where the arcs intersect.

Step #3—Construct a line from point (c) through the given point on the straight line. It will be at right angles to the original straight line.

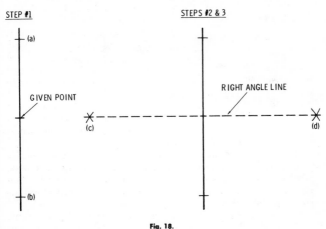

Fig. 18.

Note

If conditions permit when swinging arcs to locate point (c), arcs locating point (d) may also be swung. This will provide a double check on the accuracy of the layout as the three points, (c), given point, and (d) should form a straight line.

3-4-5 Layout Method

The 3-4-5 layout method is based on the fact that any triangle having sides with a 3-4-5 length ratio is a right triangle.

Step #1—Select a suitable measuring unit. The largest practical, as the larger the layout the less effect minor errors will have. Measure three (3) units from the given point at approximately a right angle from the original line and swing arc (a).

Step #2—Measure four (4) units from the given point along the original line and locate point (b).

Step #3—Measure five units from point (b) and locate point (c) on the arc line (a).

Step #4—Construct a line from point (c) through the given point. It will be at right angles to the original line.

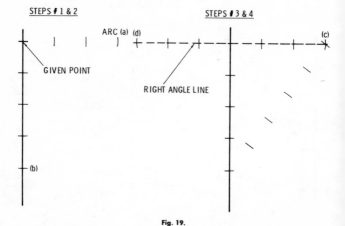

Fig. 19.

Note

If conditions permit, a similar triangle may be constructed on the opposite side of the original line. This will provide a double check on the accuracy of the layout as the three points (d), given point, and (c) should form a straight line.

In cases where layout lines are also to be used as reference or alignment lines for machinery and equipment installation, it is recommended that piano wire lines be used. High tensile strength piano wire line is superior to fiber line because it will not stretch, loosen, or sag. When properly tightened it will stay taut and relatively stationary in space, retaining its setting and providing a high degree of accuracy.

Other lines and/or points may be precisely located from such a line by measurement, use of plumb lines, etc. A fixed-point location may be easily marked on a wire line by crimping a small particle of lead or other soft material to the wire.

To make a wire line taut it must be drawn up very tightly and placed under high tensile load. To accomplish this, and to hold the line in this condition, requires rigid fastenings at the line ends and a means of tightening and securing the line. Provisions must also be

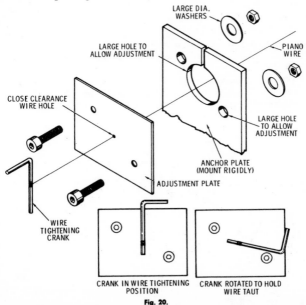

Fig. 20.

made for adjustment if the accuracy of setting, which is the principal advantage of a wire line, is to be accomplished.

Illustrated in Fig. 20 is a device which incorporates provisions for holding, tightening, adjusting and securing a piano wire line. The "anchor plate" must be rigidly mounted to maintain the high tensile load in the line. Precise adjustment is accomplished by lightly tapping on the "adjusting plate" to move it in the direction required. When making adjustments the "clamping screws" should be only lightly tightened as the adjusting plate must move on the surface of the anchor plate. When the line is accurately positioned the clamping screws should be securely tightened.

MACHINERY INSTALLATION

The first step in machinery installation is to provide a suitable base or support, termed a "foundation." It must be capable of carrying the applied load without settlement or crushing. Heavy machinery foundations are usually concrete type structures, structural steel being used for lighter applications and where space and economy are determining factors.

Anchor bolts are used to secure machinery rigidly to concrete foundations. The anchor bolts are usually equipped with a hook or some other form of fastening device to insure unity with the foundation con-

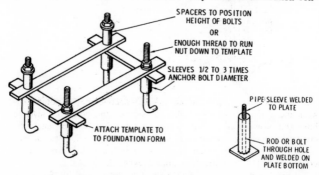

SPACERS TO POSITION
HEIGHT OF BOLTS
OR
ENOUGH THREAD TO RUN
NUT DOWN TO TEMPLATE

SLEEVES 1/2 TO 3 TIMES
ANCHOR BOLT DIAMETER

PIPE SLEEVE WELDED
TO PLATE

ATTACH TEMPLATE TO
TO FOUNDATION FORM

ROD OR BOLT
THROUGH HOLE
AND WELDED ON
PLATE BOTTOM

Fig. 21.

crete. The most common method of locating anchor bolts is making a template with holes corresponding to those in the machine to be fastened, as shown in Fig. 21. A simplified anchor bolt assembly made from readily available parts is shown in Fig. 21.

Machine units should not be mounted directly on a concrete foundation and secured with anchor bolts. Such a practice presents many difficult problems with units requiring alignment. Good design incorporates a bed plate, secured to the foundation with anchor bolts, upon which machine units are mounted and fastened. The bed plate provides a solid level supporting surface, allowing ease and accuracy of shimming and alignment. As the mounting bolts for the machinery are separate from the bed-plate anchor bolts, the bed plate is not disturbed as machine mounting bolts are loosened and tightened during alignment.

Because it is impractical to mount a bed plate directly on a concrete foundation, shims and grout are used. The shims provide a means of leveling the bed plate at the required elevation, the grout provides support and securely holds all parts in position. This method of foundation mounting is shown in the cross-section illustration of Fig. 22.

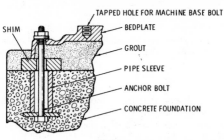

TAPPED HOLE FOR MACHINE BASE BOLT
SHIM
BEDPLATE
GROUT
PIPE SLEEVE
ANCHOR BOLT
CONCRETE FOUNDATION

Fig. 22.

Grout is a fluid mixture of mortar-like concrete that is poured between the foundation and bed plate. It secures and holds the leveling shims and provides an intimate support surface for the bed plate. The foundation should be constructed with an elevation allowance of ¾″ to 1½″ for grout. Less than ¾″ grout may crack and break up in service, while 1½″ is a practical shim thickness limit.

The number and location of shims will be determined by the design of the bed plate. Firm support should be provided at points where weight will be concentrated and at anchor bolt locations. Flat shims of the heaviest possible thickness are recommended. The use of wedge-type shims simplifies the leveling of the bed plate, however, this must be done properly to give adequate support. Wedge shims should be double and placed in line with the bottom surface of the bed plate edge as shown in Fig. 23A. The improper practice of using single wedges at right angles to the bed plate edge is shown in Fig. 23B. As only a relatively small area of the single shim is supporting the bed plate, unit stress may be excessive and crushing may occur.

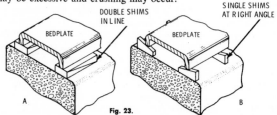

Fig. 23.

Leveling

The term "leveling", in respect to machinery installation is the operation of placing machinery or equipment on a true horizontal plane. The tool used, called a "level", is an instrument incorporating a glass tube containing spirits. A bubble is formed in the tube when it is slightly less than completely filled with the spirits. The tube is then mounted in the body of the level and calibrated to indicate true horizontal position when the bubble is centered. The accuracy of a level may be checked by comparison of the bubble readings when placed on a horizontal plane, and then reversed end-for-end. The bubble readings should be identical in both positions.

When leveling machinery the level should be used as a measuring instrument, not as a checking device for trial and error adjustments. The level should be used in conjunction with a feeler or thickness gauge to measure the amount of error or off level. The leveling shim thickness can then be calculated from this measurement.

Determining Shim Thickness

1. Insert thickness gauge as needed at low end of level to get zero bubble reading.

2. The shim thickness required will be as many times thicker than the gauge thickness as the distance between bearing points is greater than the level length. See below:

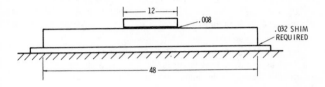

Fig. 24.

The shim thickness is 4 times the gauge thickness because the distance between bearing points is 4 times greater than the level length. The shim thickness is 4 times .008 or .032.

Elevation

In addition to the need for machinery to be installed level, in most cases it must also be placed at a given height in respect to other objects or surfaces. The usual procedure followed is to construct the foundation with an allowance for shimming. To attain a specific elevation and a level position the machine should first be roughly shimmed to approximate level at slightly below the desired elevation. One bearing point can then be shimmed to the given elevation. As this point will be high, all other bearing points can then be raised to correct elevation by shimming them level.

MACHINE ASSEMBLY

Allowance For Fits

The American Standards Association (A.S.A.) classifies machine fits into eight groups, specified as Class #1 through Class #8. The standard specifies the limits for internal and external members for different sizes in each class. The groups below, listed by common shop terms, compare approximately to the A.S.A. classes as follows:

Running Fit—A.S.A. Class #2—This fit is for assemblies where one part will run in another under load with lubrication.

*Push Fit—A.S.A. Class #4 & #5—*This fit ranges from the closest fit that can be assembled by hand, through zero clearance, to very slight interference. Assembly is selective and not interchangeable.

*Drive Fit—A.S.A. Class #6—*This fit is used where parts are to be tightly assembled and not normally disassembled. It is an interference fit and requires light pressure. It is also used as a shrink fit on light sections.

*Force or Shrink Fit—A.S.A. Class #8—*This fit requires heavy force for cold assembly or heat to assemble parts as a shrink fit. It is used where the metal can be highly stressed without exceeding its elastic limit.

Table 1. Recommended Allowances

Diameter Inches	RUNNING FIT		PUSH FIT	
	Ordinary Loads	Severe Loads	Light Service	No Play
	(Clear)	(Clear)	(Clear)	(Inter)
Up to 1/2	.0005	.001	.00025	.0000
1/2 to 1	.001	.0015	.0003	.00025
1 to 2	.002	.0025	.0003	.00025
2 to 3 1/2	.0025	.0035	.0003	.0003
3 1/2 to 6	.0035	.0045	.0005	.0005

Diameter Inches	DRIVE FIT		FORCE or SHRINK FIT	
	Field Assembly	Shop Assembly	Force	Shrink
	(Inter)	(Inter)	(Inter)	(Inter)
Up to 1/2	.0002	.0005	.00075	.001
1/2 to 1	.0002	.0005	.001	.002
1 to 2	.0005	.0008	.002	.003
2 to 3 1/2	.0005	.001	.003	.004
3 1/2 to 6	.0005	.001	.004	.005

(Clear) Indicates clearance between members
(Inter) Indicates interference between members

Keys, Key Seats, and Keyways

A key is a piece of metal placed so that part of it lies in a groove, called a *key seat* cut in a shaft. The key then extends somewhat above the shaft and fits into a *keyway* cut in a hub.

The simplest key is the square key, placed half in the shaft and half in the hub. A flat key is rectangular in cross section and is used in the same manner as the square key for members with light sections. The gib head key is tapered on its upper surface and is driven in to form a very secure fastening.

A variation on the square key is the *Woodruff key*. It is a flat disc made in the shape of a segment of a circle. It is flat on top with a round bottom to match a semicylindrical keyseat.

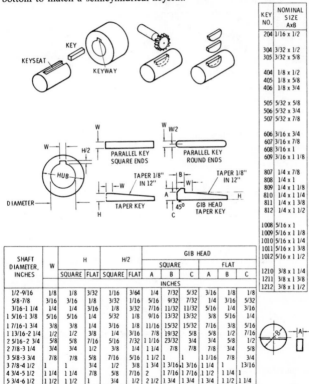

KEY NO.	NOMINAL SIZE AxB
204	1/16 x 1/2
304	3/32 x 1/2
305	3/32 x 5/8
404	1/8 x 1/2
405	1/8 x 5/8
406	1/8 x 3/4
505	5/32 x 5/8
506	5/32 x 3/4
507	5/32 x 7/8
606	3/16 x 3/4
607	3/16 x 7/8
608	3/16 x 1
609	3/16 x 1 1/8
807	1/4 x 7/8
808	1/4 x 1
809	1/4 x 1 1/8
810	1/4 x 1 1/4
811	1/4 x 1 3/8
812	1/4 x 1 1/2
1008	5/16 x 1
1009	5/16 x 1 1/8
1010	5/16 x 1 1/4
1011	5/16 x 1 3/8
1012	5/16 x 1 1/2
1210	3/8 x 1 1/4
1211	3/8 x 1 3/8
1212	3/8 x 1 1/2

SHAFT DIAMETER, INCHES	W	H		H/2		GIB HEAD					
						SQUARE			FLAT		
		SQUARE	FLAT	SQUARE	FLAT	A	B	C	A	B	C
						INCHES					
1/2-9/16	1/8	1/8	3/32	1/16	3/64	1/4	7/32	5/32	3/16	1/8	1/8
5/8-7/8	3/16	3/16	1/8	3/32	1/16	5/16	9/32	7/32	1/4	3/16	5/32
3/16-1 1/4	1/4	1/4	3/16	1/8	3/32	7/16	11/32	11/32	5/16	1/4	3/16
1 5/16-1 3/8	5/16	5/16	1/4	5/32	1/8	9/16	13/32	13/32	3/8	5/16	1/4
1 7/16-1 3/4	3/8	3/8	1/4	3/16	1/8	11/16	15/32	15/32	7/16	3/8	5/16
1 13/16-2 1/4	1/2	1/2	3/8	1/4	3/16	7/8	19/32	5/8	1/2	1/2	7/16
2 5/16-2 3/4	5/8	5/8	7/16	5/16	7/32	1 1/16	23/32	3/4	3/4	5/8	1/2
2 7/8-3 1/4	3/4	3/4	1/2	3/8	1/4	1 1/4	7/8	7/8	7/8	3/4	5/8
3 5/8-3 3/4	7/8	7/8	5/8	7/16	5/16	1 1/2	1	1	1 1/16	7/8	3/4
3 7/8-4 1/2	1	1	3/4	1/2	3/8	1 3/4	1 3/16	1 3/16	1 1/4	1	13/16
4 3/4-5 1/2	1 1/4	1 1/4	7/8	5/8	7/16	2	1 7/16	1 7/16	1 1/2	1	1
5 3/4-6 1/2	1 1/2	1 1/2	1	3/4	1/2	2 1/2	1 3/4	1 3/4	1 3/4	1 1/2	1 1/4

Fig. 25.

Scraping

In metal working, slight errors in plane or curved surfaces are often corrected by hand scraping. Most machine surfaces that slide on one another, as in machine tools, are finished in this manner. Also plane bearing boxes are scraped to fit their shafts after having been bored, or in the case of babbitt bearings, after having been poured.

Various styles of hand scrapers are shown in Fig. 26. The flat scraper (A) is used for flat scraping. The hook scraper (B) also used on flat surfaces is preferred by some. Flat and curved scrapers with a half-round cross section (C) and (D) are used for scraping bearings. The three-cornered scraper (E) is used to some extent on curved surfaces and to remove burrs and round the corners of holes.

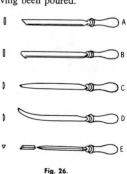

Fig. 26.

Flat Surface Scraping

Coat the entire surface of a true surface plate with a suitable scraping dye such as Prussian blue.

Place the surface plate on the surface to be scraped, or if the work piece is small, place it on the surface plate.

Move the plate or the piece back and forth a few times to color the high points on the work piece.

Scrape the high spots on the work piece which are colored with the dye where they contacted the surface plate.

Bearing Scraping

Coat the journal of the shaft with a thin layer of Prussian blue, spreading it with the forefinger.

Place the shaft in the bearing, or vice versa, tighten it, and turn one or the other several times through a small angle.

Scrape the high spots in the bearing which are colored where they contacted the shaft journal.

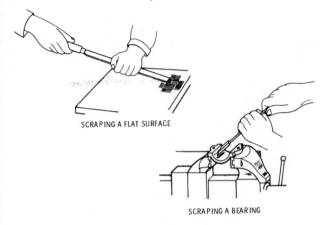

SCRAPING A FLAT SURFACE

SCRAPING A BEARING

Fig. 27.

TORQUE WRENCH

The words "torque wrench" are commonly used to describe a tool which is a combination wrench and measuring tool. It is used to apply a twisting force, as do conventional wrenches, and to simultaneously measure the magnitude of the force. This twisting force, which tends to turn a body about an axis of rotation, is called *torque*.

There are numerous types of torque wrenches, some that are direct reading, others with signaling mechanisms to warn when the predetermined torque is reached. All are based on the fundamental law of the lever; that is: *force times distance equals torque.*

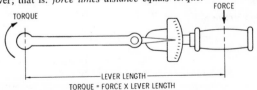

TORQUE = FORCE X LEVER LENGTH

Fig. 28.

Units of Measure

Torque units of measure (inch pound and foot pound) are the product of a force measured in pound units and a lever length measured in either inch units or foot units.

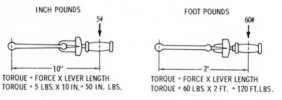

INCH POUNDS   FOOT POUNDS

TORQUE = FORCE X LEVER LENGTH   TORQUE = FORCE X LEVER LENGTH
TORQUE = 5 LBS. X 10 IN. = 50 IN. LBS.   TORQUE = 60 LBS. X 2 FT. = 120 FT. LBS.

Fig. 29.

To convert inch pounds to foot pounds, divide by 12.
Example: 60 in.lbs. equals 5 ft.lbs. (60 divided by 12)
To convert foot pounds to inch pounds, multiply by 12.
Example: 12 ft.lbs. equals 144 in.lbs. (12 multiplied by 12)

TORQUE WRENCH TERMS

Push or Pull

Force should be applied to a torque wrench by pulling whenever possible. This is primarily because there is greater hazard to fingers or knuckles when pushing should some part fail unexpectedly. While pulling is the preferred method, either way can produce accurate results.

Break-Away Torque

The torque required to loosen a fastener is generally some value lower than that to which it has been tightened. For a given size and type of fastener there is a direct relationship between tightening torque and breakaway torque. When this relationship has been determined by actual test, tightening torque may be checked by loosening and checking breakaway torque.

Set or Seizure

In the last stages of rotation in reaching a final torque reading, seizing or set of the fastener may occur. When this occurs there is a noticeable popping effect. To break the set, back off and then again apply the tightening torque. Accurate torque setting can not be made if the fastener is seized.

Run Down Resistance

The torque required to rotate a fastener before makeup occurs is a measure of run-down resistance. To obtain the proper torque value where tight threads on locknuts produce a run-down resistance, add the resistance to the required torque value. Run-down resistance must be measured on the last rotation or as close to the makeup point as possible.

Wrench Size

The correct size wrench for a job is one that will read between 25% to 75% of the scale when the required torque is applied. This will allow adequate capacity and provided satisfactory accuracy. Avoid using an oversize torque wrench, obtaining correct readings as the pointer starts up the scale is difficult. Too small a wrench will not allow for extra capacity in the event of seizure or run-down resistance.

Torque and Tension

Torque and tension are distinctively different and must not be confused. Torque is twist, the standard unit of measure being foot-pounds; tension is straight pull, the unit of measure being pounds. Wrenches designed for measuring the tightness of a threaded fastener are distinctively torque wrenches and not tension wrenches.

Attachments

Many styles of attachments are available to fit various fasteners and to reach applications that may otherwise be impossible to torque. Most of these increase the wrench capacity as they lengthen the lever arm. Therefore, when using such attachments scale reading corrections must be made. The scale correction will be in reverse ratio to the increase in lever arm length. If the arm is doubled by adding an attachment to the wrench, its capacity is doubled and the scale shows only one half of the actual torque applied. The following formula can be used to determine correct scale readings when using an attachment.

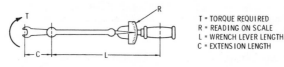

T = TORQUE REQUIRED
R = READING ON SCALE
L = WRENCH LEVER LENGTH
C = EXTENSION LENGTH

Fig. 30.

$$\text{Scale Reading} = \frac{\text{torque required} \times \text{wrench length}}{(\text{wrench length} + \text{attachment length})}$$

$$R = \frac{T \times L}{(L + C)}$$

TORQUE SPECIFICATIONS

The suggested maximum torque values in Table 2 for fasteners of various materials should be used as a guide only. Manufacturer's specifications should be followed on specific torque applications.

Table 2. Torque in Foot Pounds

Fastener Diameter	Threads Per Inch	Mild Steel	Stainless Steel 18-8	Alloy Steel
1/4	20	4	6	8
5/16	18	8	11	16
3/8	16	12	18	24
7/16	14	20	32	40
1/2	13	30	43	60
5/8	11	60	92	120
3/4	10	100	128	200
7/8	9	160	180	320
1	8	245	285	490

Torque Values for Steel Fasteners

The strength of a bolted connection depends on the clamping force developed by the bolts. The tighter the bolt, the stronger the connection. The two principle factors that limit the clamping force the bolt may develop are the bolt size and its tensile strength.

The tensile strength of a bolt depends principally on the material from which it is made. Bolt manufacturers identify bolt materials by head markings which conform to SAE and ASTM specifications. These marks are known as grade markings. The most commonly specified grades are listed in Table 3.

Table 3. Suggested Torque Values for Graded Steel Bolts

Grade		SAE 1&2	SAE 5	SAE 6	SAE 8
Tensile Strength		64000 PSI	105000 PSI	130000 PSI	150000 PSI
Grade Mark					
Bolt Dia.	Thds. Per In.	Foot Pounds Torque			
1/4	20	5	7	10	10
5/16	18	9	14	19	22
3/8	16	15	25	34	37
7/16	14	24	40	55	60
1/2	13	37	60	85	92
9/16	12	53	88	120	132
5/8	11	74	120	169	180
3/4	10	120	200	280	296
7/8	9	190	302	440	473
1	8	282	466	660	714

The above values are based on approximately 75% of yield strength.
Fasteners must be lubricated (Petroleum Lubricant)

The values in Table 3 do not apply if special lubricants such as colloidal copper or molybdenum disulphite are used. Use of special lubricants can reduce the amount of friction in the fastener assembly so the torque applied may produce far greater tension than desired.

MECHANICAL POWER TRANSMISSION

The three principal systems used for the transmission of rotary mechanical power between adjacent shafts are: belts, chains, and gears. Understanding of certain terms and concepts that are basic to all three systems is a prerequisite to understanding of the systems.

Pitch

A word commonly used in connection with machinery and mechanical operations meaning: *The distance from a point to a corresponding point.*

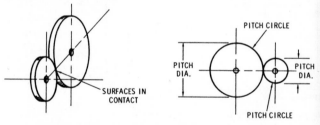

Pitch Diameter and Pitch Circle

The *pitch diameter* specifies the distance across the center of the *pitch circle*. Pitch diameter dimensions are specific values even though the *pitch circle* is imaginary. Rotary power transmission calculations are based on the concept of circles or cylinders in contact. These circles are called *pitch circles*.

As the shafts rotate, the surfaces of the pitch circles travel equal distances at equal speeds (assuming no slippage). Shafts then rotate at speeds proportional to the circumference of the pitch circles and therefore proportional to the pitch diameters.

This concept of pitch circles in contact applies to belt and chain drives although the pitch circles are actually separated. This is true because the belt or chain is in effect an extension of the pitch circle surface.

A. As rotation occurs the pitch circle surfaces will travel the same distance at the same surface speed.
B. Because the circumference of a 4″ pitch circle is double that of a 2″ pitch circle its rotation will be one half as much.

C. The rotation speed therefore of the 4″ pitch circle will be one half that of the 2″ pitch circle.

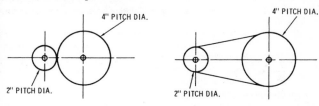

Calculations

Rotational speed and pitch diameter calculations for belts, chains, and gears are based on the concept of pitch circles in contact. The relationship that results from this concept may be stated as follows:

Shaft speeds are inversely proportional to pitch diameters.

In terms of rotational speeds and pitch diameters this relationship may be expressed in equation form as follows;

$$\frac{Driver\ Rotational\ Speed}{Driven\ Rotational\ Speed} = \frac{Driven\ Pitch\ Diameter}{Driver\ Pitch\ Diameter}$$

To simplify the use of the equation, letters and numbers instead of words, are used to represent the terms.

S1 for *DRIVER ROTATIONAL SPEED*
S2 for *DRIVEN ROTATIONAL SPEED*
P1 for *DRIVER PITCH DIAMETER*
P2 for *DRIVEN PITCH DIAMETER*

The basic equation then becomes $\dfrac{S1}{S2} = \dfrac{P2}{P1}$

The equation may be rearranged into the following forms, one for each of the four values. To find an unknown value the known values are substituted in the appropriate equation.

$$S1 = \frac{P2 \times S2}{P1} \qquad S2 = \frac{P1 \times S1}{P2} \qquad P1 = \frac{S2 \times P2}{S1} \qquad P2 = \frac{S1 \times P1}{S2}$$

The above equations may also be stated as rules. Following these rules is another convenient way to calculate unknown shaft speeds and pitch diameters.

To Find—*Driving Shaft Speed*

> Multiply *driving pitch diameter* by speed of *driven shaft* and divide by *driving pitch diameter*.

To Find—*Driven Shaft Speed*

> Multiply *driving pitch diameter* by speed of *driving shaft* and divide by speed of *driven pitch diameter*.

To Find—*Driving Pitch Diameter*

> Multiply *driven pitch diameter* by speed of *driven shaft* and divide by speed of *driving shaft*.

To Find—*Driven Pitch Diameter*

> Multiply *driving pitch diameter* by speed of *driving shaft* and divide by speed of *driven shaft*.

In gear and sprocket calculations the number of teeth is used rather than the pitch diameters. The equations and rules hold true because the number of teeth in a gear or sprocket is directly proportional to its pitch diameter.

Calculations (Examples)

To find an unknown value, substitute the known values in the appropriate equation or follow the appropriate rule.

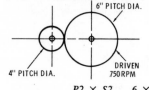

To Find Driving Shaft Speed (S1)
Known Values

> *P1* Driver Pitch Dia. = 4″
> *P2* Driven Pitch Dia. = 6″
> *S2* Driven Shaft Speed = 750 rpm

$$S1 = \frac{P2 \times S2}{P1} \quad \text{or} \quad \frac{6 \times 750}{4} \quad \text{or} \quad \frac{4500}{4} \quad \text{or } 1125 \text{ rpm}$$

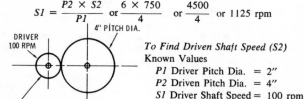

To Find Driven Shaft Speed (S2)
Known Values

> *P1* Driver Pitch Dia. = 2″
> *P2* Driven Pitch Dia. = 4″
> *S1* Driver Shaft Speed = 100 rpm

$$S2 = \frac{P1 \times S1}{P2} \quad \text{or} \quad \frac{2 \times 100}{4} \quad \text{or} \quad \frac{200}{4} \quad \text{or } 50 \text{ rpm}$$

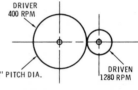

To Find Driver Pitch Dia. (P1)
Known Values
   P2 Driver Pitch Dia.   = 3"
   S1 Driver Shaft Speed = 600
   S2 Driven Shaft Speed = 2000 rpm

$$P1 = \frac{S2 \times P2}{S1} \quad \text{or} \quad \frac{2000 \times 3}{600} \quad \text{or} \quad \frac{6000}{600} \quad \text{or } 10'' \text{ pitch dia.}$$

To Find Driven Pitch Dia. (P2)
Known Values
   P1 Driver Pitch Dia.   = 8"
   S1 Driver Shaft Speed = 400
   S2 Driven Shaft Speed = 1280 rpm

$$P2 = \frac{S1 \times P1}{S2} \quad \text{or} \quad \frac{400 \times 8}{1280} \quad \text{or} \quad \frac{3200}{1280} \quad \text{or } 2.5'' \text{ pitch dia.}$$

"V" BELTS

The "V" belt has a tapered cross-sectional shape which causes it to wedge firmly into the sheave groove under load. Its driving action takes place through frictional contact between the sides of the belt and the sheave groove surfaces. While the cross-sectional shape varies slightly with make, type, and size, the included angle of most "V" belts is about 42 degrees. There are three general classifications of "V" belts: Fractional Horsepower, Standard Multiple, and Wedge.

Fractional Horsepower

Used principally as single belts on fractional horsepower drives. Designed for intermittent and relatively light loads. Manufactured in four standard cross-sectional sizes as shown in Fig. 31A.

Standard belt lengths vary by one-inch increments between a minimum length of 10 inches and a maximum length of 100 inches. In addition some fractional horsepower belts are made to fractional inch lengths.

The numbering system used indicates the cross-sectional size and the nominal outside length. The last digit of the belt number indicating tenths of an inch. Because the belt number indicates length along the outside surface, belts are slightly shorter along the pitch line than the nominal size number indicates.

Examples

Belt Number	Size	Outside Length
3L 470	3L	47″
4L 425	4L	42½″

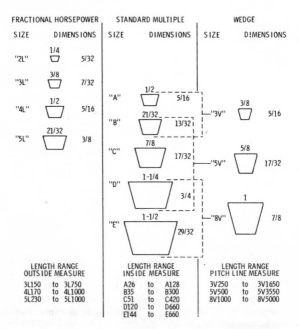

FRACTIONAL HORSEPOWER	STANDARD MULTIPLE	WEDGE

LENGTH RANGE OUTSIDE MEASURE

3L150 to 3L750
4L170 to 4L1000
5L230 to 5L1000

LENGTH RANGE INSIDE MEASURE

A26 to A128
B35 to B300
C51 to C420
D120 to D660
E144 to E660

LENGTH RANGE PITCH LINE MEASURE

3V250 to 3V1650
5V500 to 5V3550
8V1000 to 8V5000

Fig. 31A.   Fig. 31B.   Fig. 31C.

Standard Multiple

Standard multiple belts are designed for the continuous service usually encountered in industrial applications. As the name indicates, more than one belt provides the required power transmission capacity. Most manufacturers furnish two grades, a standard and a premium quality. The standard belt is suitable for the majority of industrial drives that have normal loads, speeds, center distances, sheave diameters, and operating conditions. The premium quality is made for drives subjected to severe loads, shock, vibration, temperatures, etc.

The standard multiple "V" belt is manufactured in five standard cross-sectional sizes designated: "A," "B," "C," "D," "E" as shown in Fig. 31B.

The actual pitch length of standard multiple belts may be from one to several inches greater than the nominal length indicated by the belt number. This is because the belt numbers indicate the length of the belt along its inside surface. As belt length calculations are in terms of belt length on the pitch line, a table of pitch line belt lengths is recommended when selecting belts.

Wedge

The wedge belt is an improved design "V" belt which makes possible a reduction in size, weight, and cost of "V" belt drives. Utilizing improved materials, these multiple belts have a smaller cross-section per horsepower and use smaller diameter sheaves at shorter center distances than is possible with standard multiple belts. Because of the premium-quality heavy-duty construction, only three cross-sectional belt sizes are used to cover the duty range of the five sizes of standard multiple belts. The dimensions of the three standard "wedge" belt cross-sectional sizes "3V," "5V," "8V" are shown in Fig. 31C.

The wedge belt number indicates the number of ⅛ inches of top width of the belt. As shown above the "3V" belt has a top width of ⅜ inches, the "5V" a width of ⅝ inches and the "8V" a full one inch of top width.

The belt length indicated by the wedge belt number is the effective pitch line length of the belt. As belt calculations are in terms of pitch line lengths, nominal belt numbers can be used directly when choosing wedge belts.

"V" BELTS

Table 4. Standard "V" Belt Lengths

A BELTS			B BELTS			C BELTS		
BELT NUMBER Standard	Pitch Length	Outside Length	BELT NUMBER Standard	Pitch Length	Outside Length	BELT NUMBER Standard	Pitch Length	Outside Length
A26	27.3	28.0	B35	36.8	38.0	C51	53.9	55.0
A31	32.3	33.0	B38	39.8	41.0	C60	62.9	64.0
A35	36.3	37.0	B42	43.8	45.0	C68	70.9	72.0
A38	39.3	40.0	B46	47.8	49.0	C75	77.9	79.0
A42	43.3	44.0	B51	52.8	54.0	C81	83.9	85.0
A46	47.3	48.0	B55	56.8	58.0	C85	87.9	89.0
A51	52.3	53.0	B60	61.8	63.0	C90	92.9	94.0
A55	56.3	57.0	B68	69.8	71.0	C96	98.9	100.0
A60	61.3	62.0	B75	76.8	78.0	C105	107.9	109.0
A68	69.3	70.0	B81	82.8	84.0	C112	114.9	116.0
A75	76.3	77.0	B85	86.8	88.0	C120	122.9	124.0
A80	81.3	82.0	B90	91.8	93.0	C128	130.9	132.0
A85	86.3	87.0	B97	98.8	100.0	C136	138.9	140.0
A90	91.3	92.0	B105	106.8	108.0	C144	146.9	148.0
A96	97.3	98.0	B112	113.8	115.0	C158	160.9	162.0
A105	106.3	107.0	B120	121.8	123.0	C162	164.9	166.0
A112	113.3	114.0	B128	129.8	131.0	C173	175.9	177.0
A120	121.3	122.0	B136	137.8	139.0	C180	182.9	184.0
A128	129.3	130.0	B144	145.8	147.0	C195	197.9	199.0
			B158	159.8	161.0	C210	212.9	214.0
			B173	174.8	176.0	C240	240.9	242.0
			B180	181.8	183.0	C270	270.9	272.0
			B195	196.8	198.0	C300	300.9	302.0
			B210	211.8	213.0	C360	360.9	362.0
			B240	240.3	241.5	C390	390.9	392.0
			B270	270.3	271.5	C420	420.9	422.0
			B300	300.3	301.5			

D BELTS			3V BELTS		5V BELTS		8V BELTS	
BELT NUMBER Standard	Pitch Length	Outside Length	Belt No.	Belt Length	Belt No.	Belt Length	Belt No.	Belt Length
D120	123.3	125.0	3V250	25.0	5V500	50.0	8V1000	100.0
D128	131.3	133.0	3V265	26.5	5V530	53.0	8V1060	106.0
D144	147.3	149.0	3V280	28.0	5V560	56.0	8V1120	112.0
D158	161.3	163.0	3V300	30.0	5V600	60.0	8V1180	118.0
D162	165.3	167.0	3V315	31.5	5V630	63.0	8V1250	125.0
D173	176.3	178.0	3V335	33.5	5V670	67.0	8V1320	132.0
D180	183.3	185.0	3V355	35.5	5V710	71.0	8V1400	140.0
D195	198.3	200.0	3V375	37.5	5V750	75.0	8V1500	150.0
			3V400	40.0	5V800	80.0	8V1600	160.0

Table 4. Standard "V" Belt Lengths (continued)

Belt Number Standard	Pitch Length	Outside Length
D210	213.3	215.0
D240	240.8	242.5
D270	270.8	272.5
D300	300.8	302.5
D330	330.8	332.5
D360	360.8	362.5
D390	390.8	392.5
D420	420.8	422.5
D480	480.8	482.5
D540	540.8	542.5
D600	600.8	602.5

E BELTS

Belt Number Standard	Pitch Length	Outside Length
E180	184.5	187.5
E195	199.5	202.5
E210	214.5	217.5
E240	241.0	244.0
E270	271.0	274.0
E300	301.0	304.0
E330	331.0	334.0
E360	361.0	364.0
E390	391.0	394.0
E420	421.0	424.0
E480	481.0	484.0
E540	541.0	544.0
E600	601.0	604.0

3V Belt	Length	5V Belt	Length	8V Belt	Length
3V425	42.5	5V850	85.0	8V1700	170.0
3V450	45.0	5V900	90.0	8V1800	180.0
3V475	47.5	5V950	95.0	8V1900	190.0
3V500	50.0	5V1000	100.0	8V2000	200.0
3V530	53.0	5V1060	106.0	8V2120	212.0
3V560	56.0	5V1120	112.0	8V2240	224.0
3V600	60.0	5V1180	118.0	8V2360	236.0
3V630	63.0	5V1250	125.0	8V2500	250.0
3V670	67.0	5V1320	132.0	8V2650	265.0
3V710	71.0	5V1400	140.0	8V2800	280.0
3V750	75.0	5V1500	150.0	8V3000	300.0
3V800	80.0	5V1600	160.0	8V3150	315.0
3V850	85.0	5V1700	170.0	8V3350	335.0
3V900	90.0	5V1800	180.0	8V3550	355.0
3V950	95.0	5V1900	190.0	8V3750	375.0
3V1000	100.0	5V2000	200.0	8V4000	400.0
3V1060	106.0	5V2120	212.0	8V4250	425.0
3V1120	112.0	5V2240	224.0	8V4500	450.0
3V1180	118.0	5V2360	236.0	*8V5000	500.0
3V1250	125.0	5V2500	250.0	....	....
3V1320	132.0	5V2650	265.0	....	....
3V1400	140.0	5V2800	280.0	....	....
....	....	5V3000	300.0	....	....
....	....	5V3150	315.0	....	....
....	....	5V3350	335.0	....	....
....	....	5V3550	355.0	....	....

"V" Belt Matching

Satisfactory operation of multiple belt drives require that each belt carry its share of the load. To accomplish this all belts in a drive must be essentially equal length. Because it is not economically practical to manufacture belts to exact length, most manufacturers follow a practice of code marking to indicate exact length.

Each belt is measured under specific tension and marked with a code number to indicate its variation from nominal length. The number 50 is commonly used as the code number to indicate a belt within tolerance of its nominal length. For each $\frac{1}{10}$ of an inch over nominal length, the number 50 in increased by 1. For each $\frac{1}{10}$ of an inch under nominal length, 1 is subtracted from the number 50. Most manufacturers code mark as shown in Fig. 32.

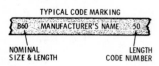

TYPICAL CODE MARKING

B60 ...MANUFACTURER'S NAME... 50

NOMINAL   LENGTH
SIZE & LENGTH   CODE NUMBER

Fig. 32.

For example, if the 60 inch "B" section belt shown above is manufactured $\frac{3}{10}$ of an inch long, it will be code marked 53 rather than the 50 shown. Or if made $\frac{3}{10}$ of an inch short it will be code marked 47. While both of these belts have the belt number B60 they cannot be used satisfactorily in a set because of the difference in their actual length.

It is possible for the length of belts to change slightly during storage. Under satisfactory conditions however, changes will not exceed measuring tolerances. Therefore, belts may be combined by matching code numbers. Ideally, sets should be made up of belts having the same code numbers, however, the resiliency of the belts allow some length variation. Table 5 lists the maximum recommended variations for standard multiple belts when making up matched belt sets.

Table 5.

Matching Number Range	Belt Lengths				
	A	B	C	D	E
2	26-180	35-180	51-180		
3		195-315	195-255	120-255	144-240
4			270-360	270-360	270-360
6			390-420	390-660	390-660

"V" Belt Drive Alignment

The life of a "V" belt is dependent on: first the quality of materials and manufacture; and second on installation and maintenance. One o the most important installation factors influencing operating life is belt alignment. In fact, excessive misalignment is probably the most frequent cause of shortened belt life.

While "V" belts because of their inherent flexibility can accommodate themselves to a degree of misalignment not tolerated by other types of power transmission, they still must be held within reasonable limits. Maximum life can be attained only with true alignment and as misalignment increases belt life is proportionally reduced. If misalignment is greater than $\frac{1}{16}$ inch for each 12 inches of center distance, very rapid wear will result.

Misalignment of belt drives results from shafts being out of angular or parallel alignment, or from the sheave grooves being out of axial alignment. These three types of misalignment are illustrated in Fig. 33.

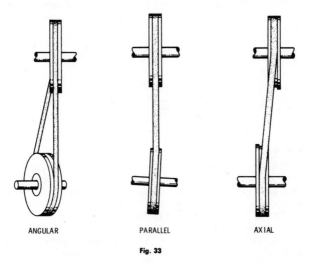

ANGULAR   PARALLEL   AXIAL

Fig. 33

Because the shafts of most "V" belt drives are in a horizontal plane, angular shaft alignment is easily obtained by leveling the shafts. In those cases where shafts are not horizontal, a careful check must be made to ensure the angle of inclination of both shafts is the same.

Before any check is made for parallel-shaft and axial-groove alignment, the shafts and sheaves must be checked for run-out. Any shaft run-out or sheave wobble will cause proportionate inaccuracies in alignment.

The checking and adjusting for parallel-shaft and axial-groove alignment of most drives can be done simultaneously if the shafts and sheaves are true.

"V" Belt Alignment

The most satisfactory method of checking parallel-shaft and axial-groove alignment is with a straightedge. It may also be done with a taut line, however, when using this method care must be exercised as the line is easily distorted. The straightedge checking method is illustrated in Fig. 34, with arrows indicating the four check points. When sheaves are properly aligned no light should be visible at these four points.

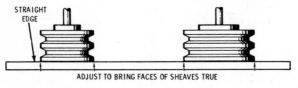

STRAIGHT EDGE

ADJUST TO BRING FACES OF SHEAVES TRUE

Fig. 34.

"V" Belt Installation

"V" belts should never be "run-on" to sheaves. To do so places excessive stress on the cords, usually straining or breaking some of them. A belt damaged in this manner will flop under load and turn over in the sheave groove. The proper installation method is to loosen the adjustable mount, reduce the center distance, and slip the belts loosely into the sheave grooves.

The following six general rules should be followed when installing "V" belts:

1. Reduce centers so belts can be slipped on sheaves.
2. Have all belts slack on the same side (top of drive).
3. Tighten belts to approximately correct tension.
4. Start unit and allow belts to seat in grooves.

5. Stop—retighten to correct tension.
6. Recheck belt tension after 24 to 48 hours of operation.

Checking Belt Tension

Belt tension is a vital factor in operating efficiency and service life. Too low a tension results in slippage and rapid wear of both belts and sheave grooves. Too high a tension stresses the belts excessively and unnecessarily increases bearing loads.

The tensioning of fractional horsepower and standard multiple belts may be done satisfactorily by tightening until the proper "feel" is attained. The proper "feel" is when the belt has a live springy action when struck with the hand. If there is insufficient tension the belt will feel loose or dead when struck. Too much tension will cause the belts to feel taut, as there will be no give to them.

Wedge Belt Tension

The 3V, 5V, and 8V wedge belts operate under very high tension since there are fewer belts and/or smaller belts per horsepower. When properly tightened they are taut and have little give, therefore, tightening by "feel" is not dependable. A better method is to use the belt-tension measuring tool shown in Fig. 35.

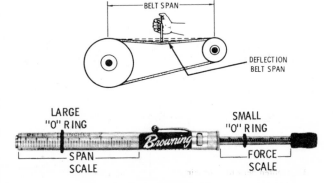

Fig. 35.

BROWNING BELT TENSION CHECKER

Instructions

To determine the lbs. force required to tension a drive with the BROWNING Belt Tensioner you simply do the following:

1. Measure the Belt Span as shown

2. Divide belt span by 64 to get belt deflection needed to check tension

3. Set large "O" ring on span scale at required belt deflection. This scale is in $\frac{1}{16}$" increments

4. Set small "O" ring at zero on the "Force Scale" (plunger)

5. Place the larger end of the tension checker squarely on one belt at the center of the belt span. Apply force on the plunger until the bottom of the large "O" ring is even with the top of the next belt or with the bottom of a straight edge laid across the sheaves

6. Read the force scale under the small "O" ring to determine the force required to give the needed deflection.

7. Compare the force scale reading with the correct value for the belt style and cross section used as given in table below. The force should be between the minimum and maximum values shown.

8. If there is too little deflection force, the belts should be tightened. If there is too much deflection force, the belts should be loosened.

Belt Cross Section	Small P.D. Range	Deflection Force Lbs.	
		Min.	Max.
3V	2.65 - 3.65	3	4½
	4.12 - 6.90	4	6
5V	7.1 - 10.9	8	12
	11.8 - 16.0	10	15
8V	12.5 - 17.0	18	27
	18.0 - 22.4	20	30

"V" Belt Replacement

When replacing "V" belts care must be exercised that the correct type is selected. Errors in choice might be made since the top width of some of the sizes in the three types are essentially the same. Also, belts from different manufacturers should not be mixed on the same drive because of variations from nominal dimensions.

When determining the length belt required for most drives it is not necessary to be exact. First, because of the adjustment built into most drives, and second because belt selection is limited to the standard lengths available. Since the standard lengths vary in steps of several inches, an approximate length calculation is usually adequate. For these reasons the following easy method of belt calculation can be used for most "V" belt drives:

1. Add the pitch diameters of the sheaves and multiply by 1½.

2. To this add twice the distance between centers.

3. Select the nearest *longer* standard belt.

This method should not be used when centers are fixed or if there are extreme pitch-diameter differences on short centers.

Sheave Groove Wear

All "V" belts and sheaves will wear to some degree with use. As wear occurs the belts will ride lower in the grooves. Generally a new belt should not seat more than 1/16 inch below the top of the groove. While belt wear is usually noticed, sheave-groove wear is often overlooked.

As wear occurs at the contact surfaces on the sides of the grooves, a dished condition develops. This results in reduced wedging action, loss of gripping power, and accelerated wear as slippage occurs. Installing new belts in worn grooves will give temporary improvement in operation but belt wear will be rapid. When changing belts therefore, Sheave-groove wear should be checked with gauges or templates.

Care must be taken when checking grooves that the correct gauge or template in respect to type, size and pitch diameter is used. As sheave grooves are designed to conform to the belt cross-section change as it bends, small diameter sheaves have less angle than larger diameter sheaves. The variation in sheave-groove included angles ranges from 34 degrees for small diameter sheaves up to 42 degrees for the largest diameter sheaves.

Sheave Grooves

When sheave groove wear becomes excessive, shoulders will develop on the groove side walls. If the sheave is not repaired or replaced these shoulders will quickly chew the bottom corners off new belts and ruin them.

Max. Allowable Wear
.025" Moderately Loaded Drives
.015" Heavily Loaded Drives

Table 6. Standard Groove Dimension Table

V-Belt	Minimum Recommended Pitch Diam.	Pitch Diam.	Angle Groove	W	D	X	S	E
A	3.0	2.6 to 5.4	34°	.494	.490	.125	5/8	3/8
		Over 5.4	38°	.504				
B	5.4	4.6 to 7.0	34°	.637	.580	.175	3/4	1/2
		Over 7.0	38°	.650				
C	9.0	7.0 to 7.99	34°	.879	.780	.200	1	11/16
		8.0 to 12.0	36°	.887				
		Over 12.0	38°	.895				
D	13.0	12.0 to 12.99	34°	1.259	1.050	.300	17/32	7/8
		13.0 to 17.0	36°	1.271				
		Over 17.0	38°	1.283				
E	21.0	18.0 to 24.0	36°	1.527	1.300	.400	1 3/4	1 1/8
		Over 24.0	38°	1.542				

FILE BREAK ALL SHARP CORNERS

GROOVE ANGLE

FACE WIDTH OF STANDARD V-BELT PULLEYS
FACE WIDTH = S(N−1)+2E
WHERE:
N = NUMBER OF GROOVES

Belt						GROOVE DIMENSIONS IN INCHES					
A	B	C	D	E	W	T	U	V	Angle of Groove	Use on O.D.
3V 11/32	13/32	.025	.325	.350	.350	.056	.123	.334	36°	under 3.5
							.109	.333	38°	3.5 to 6.0
							.096	.332	40°	6.01 to 12.0
							.081	.331	42°	12.01, over
5V 1/2	11/16	.05	.550	.600	.600	.0875	.187	.566	38°	under 10.0
							.163	.564	40°	10.1 to 16.0
							.139	.562	42°	16.01, over
8V 3/4	1 1/8	.10	.900	1.000	1.000	.125	.312	.931	38°	under 16.0
							.272	.927	40°	16.0 to 22.4
							.232	.923	42°	22.41, over

Fig. 36.

The more heavily loaded a drive, the greater the effect of groove wear on its operation. Light to moderately loaded drives may tolerate as much as ⅟₃₂" wear, whereas, heavily loaded drives will be adversely affected by .010" to .015" of wear. Wear should be checked with the appropriate gauge at the point illustrated in Fig. 36.

CHAIN DRIVES

A chain drive consists of an endless chain whose links mesh with toothed wheels called sprockets. Chain drives maintain a positive ratio between the driving and driven shafts, as they transmit power without slip or creep. The *roller chain* is the most widely used of the various styles of power transmission chain.

Roller Chain

Roller chain is composed of an alternating series of "roller links" and "pin links." The roller links consist of two pin-link plates, two bushings and two rollers. The rollers turn freely on the bushings which are press-fitted into the link plates. The pin links consist of two link plates into which two pins are press-fitted.

In operation the pins move freely inside the bushings while the rollers turn on the outside of the bushings. The relationship of the pins, bushings, rollers, and link plates is illustrated in Fig. 37.

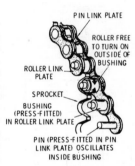

PIN LINK PLATE

ROLLER FREE TO TURN ON OUTSIDE OF BUSHING

ROLLER LINK PLATE

SPROCKET

BUSHING (PRESS-FITTED) IN ROLLER LINK PLATE

PIN (PRESS-FITTED IN PIN LINK PLATE) OSCILLATES INSIDE BUSHING

Fig. 37.

Roller Chain Dimensions

The principal roller chain dimensions are "pitch," "chain width," and "roller diameter." These dimensions are standardized, and although there are slight differences between manufacturer's products, because of this standardization chains and sprockets of different manufacturers are interchangeable.

"Pitch" is the distance from a point on one link to a corresponding point on an adjacent link. "Chain width" is the minimum distance between link plates of a roller link. "Roller diameter" is the outside diameter of a roller, and is approximately ⅝ of the pitch.

Standard series chains range in pitch from ¼ inch to 3 inches. There is also a heavy series chain ranging from ¾-inch to 3-inch pitch. The heavy series dimensions are the same as the standard except that the heavy series has thicker link plates.

Standard Roller Chain Numbers

The standard roller chain numbering system provides complete identification of a chain by number. The right hand digit in the chain

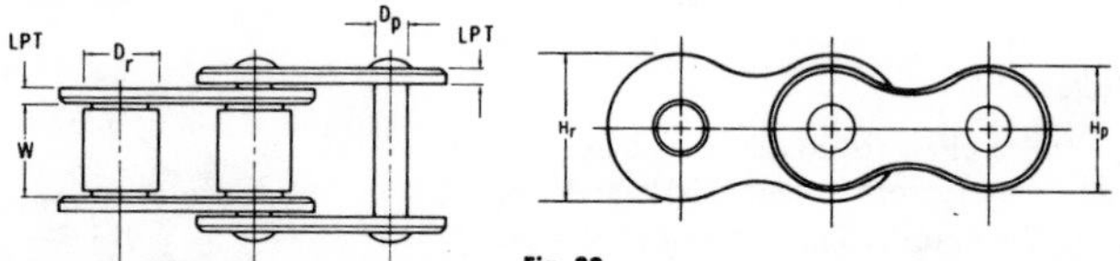

Fig. 38.

Table 7. American Standard Roller Chain Dimensions

USA STANDARD CHAIN NO. STD.	USA STANDARD CHAIN NO. HEAVY	PITCH (P)	MAX ROLLER DIAMETER (D_R)	(WIDTH) (W)	PIN DIAMETER (D_P)	THICKNESS LINK PLATE (LPT) STD.	THICKNESS LINK PLATE (LPT) HEAVY
25*	—	¼	0.130*	⅛	0.0905	0.030	—
35*	—	⅜	0.200*	3/16	0.141	0.050	—
41†	—	½	0.306	¼	0.141	0.050	—
40	—	½	5/16	5/16	0.156	0.060	—
50	—	⅝	0.400	⅜	0.200	0.080	—
60	60H	¾	15/32	½	0.234	0.094	.125
80	80H	1	⅝	⅝	0.312	0.125	.156
100	100H	1¼	¾	¾	0.375	0.156	.187
120	120H	1½	⅞	1	0.437	0.187	.219
140	140H	1¾	1	1	0.500	0.219	.250
160	160H	2	1⅛	1¼	0.562	0.250	.281
180	180H	2¼	1 13/32	1 13/32	0.687	0.281	.312
200	200H	2½	1 9/16	1½	0.781	0.312	.375
240	240H	3	1⅞	1⅞	0.937	0.375	.500

* Without rollers.

†Lightweight Chain.

number is 0 for chain of the usual proportions, 1 for lightweight chain, and 5 for a rollerless bushing chain. The number to the left of the right-hand figure, denotes the number of ⅛ inches in the pitch. The letter H following the chain number denotes the heavy series. The hyphenated 2 suffixed to the chain number denotes a double strand chain, 3 a triple strand, etc.

For example, the number 60 indicates a chain with 6, ⅛ th's, or ¾-inch, pitch. The number 41 a narrow lightweight ½-inch pitch chain. The number 25 indicates a ¼-inch pitch rollerless chain. The number 120 a chain having 12, ⅛ th's, or 1½ inches, pitch. In multiple strand chains 50-2 designates two strands of 50 chain, 50-3 triple strand, etc. General chain dimensions for standard roller chain from ¼-inch pitch to 3-inch pitch are tabulated in Table 7.

Roller Chain Connections

A length of roller chain before it is made endless will normally be made up of an even number of pitches. At either end will be an unconnected roller link with an open bushing. A special type of pin link called a *connecting link* is used to connect the two ends. The partially assembled connecting link consists of two pins press-fitted and riveted in one link plate. The pin holes in the free link plate are sized for either a slip fit or a light press fit on the exposed pins. The plate is secured in place either by cotter pin as shown in Fig. 39A, or by a spring clip as shown in Fig. 39B.

A   B

Fig. 39.

If an odd number of pitches is required an *offset link* may be substituted for an end roller link. A more stable method of providing an odd number of pitches is by use of the *offset section*. A standard connecting link is used with the offset section (Fig. 40).

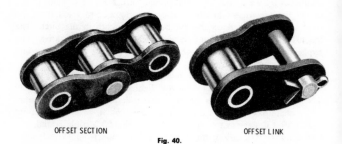

OFFSET SECTION   OFFSET LINK

Fig. 40.

Roller Chain Sprockets

Roller chain sprockets are made to standard dimensions, tolerances and tooth form. The standard includes four types (Fig. 41.); type A a plain sprocket without hubs; type B has a hub on one side; type C has a hub on both sides; type D has a detachable hub.

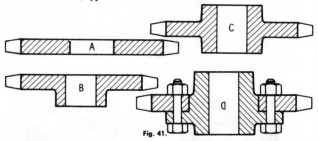

Fig. 41.

Roller Chain Installation

Correct installation of a roller chain drive requires that the shafts and the sprockets be accurately aligned. Shafts must be set level, or if inclined from a level position, both shafts must be at exactly the same angle. The shafts must also be positioned parallel within very close limits. The sprockets must be in true axial alignment for correct sprocket tooth and chain alignment.

Horizontal shafts may be aligned with the aid of a spirit level. The bubble in the level will tell when they are both in exact horizontal position. Shafts may be adjusted for parallel alignment as shown in Fig. 42. Any suitable measuring device such as calipers, feeler bars, etc., may be used. The distance between shafts on both sides of the sprockets should be equal. For an adjustable shaft drive make the distance less than final operating distance for easier chain installation. For drives with fixed shafts, the center distance must be set at the exact dimension specified.

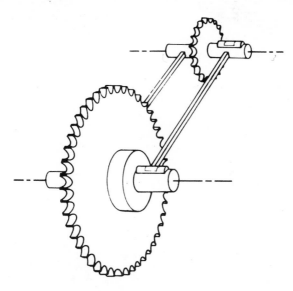

Fig. 42.

To set axial alignment of the sprockets, apply a straight-edge to the machined side surfaces as shown in Fig. 43. Tighten the set screws in the hubs to hold the sprockets and keys in position. If one of the sprockets is subject to end float, locate the sprocket so that it will be aligned when the shaft is in its normal running position. If the center distance is too great for the available straightedge, a taut piano wire may be used.

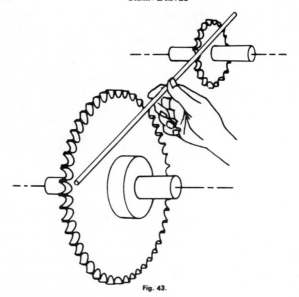

Fig. 43.

Roller Chain Installation

Fig. 44.

To install roller chain, fit it on both sprockets, bringing the free chain ends together on one sprocket. Insert the pins of the connecting link in the two end links of the chain as shown in Fig. 44; then install the free plate of the connecting link. Fasten the plate with the cotters or spring clip depending on type used. When fastened, tap back the ends of the connecting link pins so the outside of the free plate comes snugly against the fastener. This will prevent the connecting link squeezing the sprocket teeth which might interfere with free flexing of the joint and proper lubrication.

Adjustable drives must be positioned to provide proper chain tension. Horizontal and inclined drives should have an initial sag equal to 2% of the shaft centers. Measurements are made as shown in Fig. 45. The table shows measurements for various center distances to obtain approximately the recommended 2% sag.

Shaft Centers	Sag
18″	⅜″
24″	½″
30″	⅝″
36″	¾″
42″	⅞″
48″	1 ″
54″	1⅛″
60″	1¼″
70″	1½″
80″	1⅝″
90″	1⅞″
100″	2 ″
125″	2½″

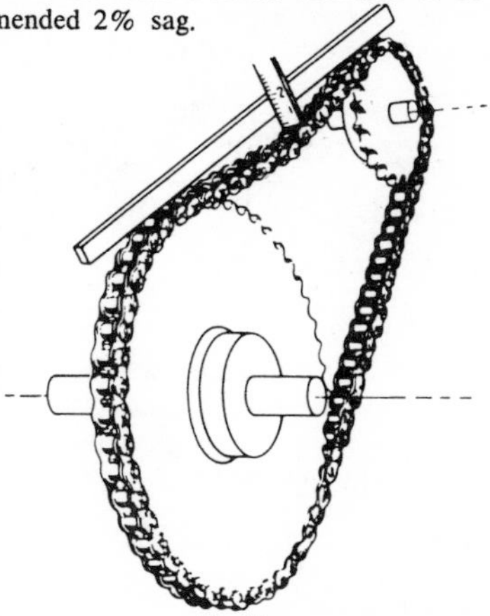

Fig. 45.

To measure the amount of sag, pull the bottom side of the chain taut so that all of the excess chain will be in the top span. Pull the top side of the chain down at its center and measure the sag as shown in Fig. 45, then adjust the centers until the proper amount is obtained. Make sure the shafts are rigidly supported and securely anchored to prevent deflection or movement which would destroy alignment.

Silent Chain

Silent chain, also called *inverted-tooth* chain, is constructed of leaf links having inverted teeth so designed that they engage cut tooth wheels in a manner similar to the way a rack engages a gear. The chain links are alternately assembled, either with pins or a combination of joint components.

Silent chain and sprockets are manufactured to a standard that is intended primarily to provide for interchangeability between chains and sprockets of different manufacturers. It does not provide for a standardization of joint components and link plate contours, which differ in each manufacturer's design. However, all manufacturer's links are contoured to engage the standard sprocket tooth, so joint centers lie on pitch diameter of the sprocket. The general proportions and designations of a typical silent chain link are shown in Fig. 46.

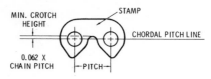

Fig. 46.

Silent chain is manufactured in a wide range of pitches and widths in various styles. Chain under ¾-inch pitch has outside guide links which engage the sides of the sprocket. The most widely used style is the *middle-guide* design with one or more rows of guide links that fit guide grooves in the sprockets. (Some manufacturers use the term "wheel" rather than sprocket).

Silent chains are designated by a combined letter and number symbol as follows:

1. A two-letter symbol: SC.
2. One or two numerical digits indicating the pitch in eighths of inches. (Usually stamped on each chain link.)
3. Two or three numerical digits indicating the chain width in quarter inches.

For example, the number "SC302" designates a silent chain of ⅜-inch pitch and ½-inch width. Or the number "SC1012" designates a silent chain of 1¼-inch pitch and 3 inches width.

Chain Replacement

During operation chain pins and bushings slide against each other as the chain engages, wraps, and disengages from the sprockets. Even when parts are well lubricated, some metal-to-metal contact does occur, and these parts eventually wear. This progressive joint wear elongates chain pitch, causing the chain to lengthen and ride higher on the sprocket teeth. The number of teeth in the large sprocket determines the amount of joint wear that can be tolerated before the chain jumps or rides over the ends of the sprocket teeth. When this critical degree of elongation is reached, the chain must be replaced.

Chain manufacturers have established tables of maximum elongation to aid in the determination of when wear has reached a critical point and replacement should be made. By placing a certain number of pitches under tension, elongation can be measured. When elongation reaches the limits recommended in the table, the chain should be replaced.

The recommended measuring procedure is to remove the chain and suspend it vertically with a weight attached to the bottom. When the chain must be measured while on sprockets, remove all slack and apply sufficient tension to keep the chain section that is being measured taut.

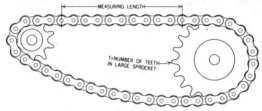

Fig. 47.

Table 8. Chain Elongation Limits

Chain Number	Pitch inches	Measuring length Number of pitches	Measuring length Nominal length inches	Length of chain when replacement is required ■ Number of teeth in largest sprocket (T)							
				Up to 67	68-73	74-81	82-90	91-103	104-118	119-140	141-173
35	3/8	32	12	$12\frac{3}{8}$	$12\frac{11}{32}$	$12\frac{5}{16}$	$12\frac{9}{32}$	$12\frac{1}{4}$	$12\frac{7}{32}$	$12\frac{3}{16}$	$12\frac{5}{32}$
40	1/2	24	12	$12\frac{3}{8}$	$12\frac{11}{32}$	$12\frac{5}{16}$	$12\frac{9}{32}$	$12\frac{1}{4}$	$12\frac{7}{32}$	$12\frac{3}{16}$	$12\frac{5}{32}$
50	5/8	20	12½	$12\frac{7}{8}$	$12\frac{15}{32}$	$12\frac{13}{16}$	$12\frac{21}{32}$	$12\frac{3}{4}$	$12\frac{25}{32}$	$12\frac{11}{16}$	$12\frac{21}{32}$
60	3/4	16	12	$12\frac{3}{8}$	$12\frac{11}{32}$	$12\frac{5}{16}$	$12\frac{9}{32}$	$12\frac{1}{4}$	$12\frac{7}{32}$	$12\frac{3}{16}$	$12\frac{5}{32}$
80	1	24	24	$24\frac{3}{4}$	$24\frac{11}{16}$	$24\frac{5}{8}$	$24\frac{9}{16}$	$24\frac{1}{2}$	$24\frac{7}{16}$	$24\frac{3}{8}$	$24\frac{5}{16}$
100	1¼	20	25	$25\frac{3}{4}$	$25\frac{11}{16}$	$25\frac{5}{8}$	$25\frac{9}{16}$	$25\frac{1}{2}$	$25\frac{7}{16}$	$25\frac{3}{8}$	$25\frac{5}{16}$
120	1½	16	24	$24\frac{3}{4}$	$24\frac{11}{16}$	$24\frac{5}{8}$	$24\frac{9}{16}$	$24\frac{1}{2}$	$24\frac{7}{16}$	$24\frac{3}{8}$	$24\frac{5}{16}$
140	1¾	14	24½	$25\frac{1}{4}$	$25\frac{3}{16}$	$25\frac{1}{8}$	$25\frac{1}{16}$	25	$24\frac{15}{16}$	$24\frac{7}{8}$	$24\frac{13}{16}$
160	2	12	24	$24\frac{3}{4}$	$24\frac{11}{16}$	$24\frac{5}{8}$	$24\frac{9}{16}$	$24\frac{1}{2}$	$24\frac{7}{16}$	$24\frac{3}{8}$	$24\frac{5}{16}$
180	2¼	11	24¾	$25\frac{1}{2}$	$25\frac{7}{16}$	$25\frac{3}{8}$	$25\frac{5}{16}$	$25\frac{1}{4}$	$25\frac{3}{16}$	$25\frac{1}{8}$	$25\frac{1}{16}$
200	2½	10	25	$25\frac{3}{4}$	$25\frac{11}{16}$	$25\frac{5}{8}$	$25\frac{9}{16}$	$25\frac{1}{2}$	$25\frac{7}{16}$	$25\frac{3}{8}$	$25\frac{5}{16}$
240	3	8	24	$24\frac{3}{4}$	$24\frac{11}{16}$	$24\frac{5}{8}$	$24\frac{9}{16}$	$24\frac{1}{2}$	$24\frac{7}{16}$	$24\frac{3}{8}$	$24\frac{5}{16}$

■ Valid for drives with adjustable centers or drives employing adjustable idler sprockets.

SPUR GEARS

The spur gear might be called the basic gear as all other types have been developed from it. Its teeth are straight and parallel to the bore center line. Spur gears may run together with other spur gears on parallel shafts; with internal gears on parallel shafts; and with a rack. A rack is in effect a straight line gear. The smaller of a pair of gears is often called a *pinion*. On large heavy duty drives the larger of a pair of gears is often called a *bullgear*.

The involute profile or form is the one commonly used for gear teeth. It is a curve that is traced by a point on the end of a taut line unwinding from a circle. The larger the circle the straighter the curvaature, and for a rack, which is essentially an infinitely large gear, the form is straight or flat.

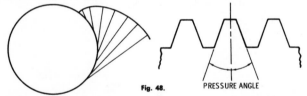

Fig. 48.   PRESSURE ANGLE

The involute system of gearing is based on a rack having straight or flat-sided teeth. The sides of each tooth incline toward the center top at an angle called the *pressure angle*. The 14½-degree pressure angle was standard for many years, however, the use of the 20-degree pressure angle has been growing until today 14½-degree gearing is generally limited to replacement work. The advantages of 20-degree gearing are greater strength and wear resistance, and in addition it permits the use of pinions with a few less teeth.

It is extremely important that the pressure angle be known when gears are mated as all gears that run together *must have the same pressure angle*.

Many types and designs of gears have been developed from the spur gear. While they are commonly used in industry, many are complex in design and manufacture. Some of the types in wide use are: bevel gears, helical gears, herringbone gears, and worm gears. Each type in turn has many specialized design variations.

Pitch Diameters and Center Distance

Pitch circles are the imaginary circles that are in contact when two standard gears are in correct mesh. The diameter of these circles are the pitch diameters of the gears. The center distance of two gears in correct mesh is equal to one half the sum of the two pitch diameters.

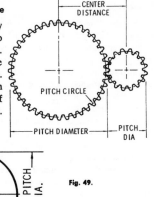

Fig. 49.

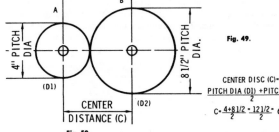

CENTER DISC (C)=
$$\frac{\text{PITCH DIA (D1) + PITCH DIA (D2)}}{2}$$
$$C = \frac{4 + 8\,1/2}{2} = \frac{12\,1/2}{2} = 6\,1/4''$$

Fig. 50.

Circular Pitch

The size and proportions of gear teeth are designated by a specific type of pitch. In gearing terms there are two specific types of pitch. They are *circular pitch* and *diametral pitch*. Circular pitch is simply the distance from a point on one tooth to a corresponding point on the next tooth, measured along the pitch line as illustrated below. Large gears are frequently made to circular pitch dimensions.

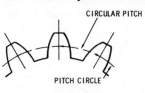

Fig. 51.

Diametral Pitch

The *diametral-pitch* system is the most widely used gearing system, practically all common size gears being made to diametral-pitch dimensions. Diametral-pitch numbers designate the size and proportions of gear teeth by specifying the number of teeth per inch of pitch diameter. For instance, a 12 diametral-pitch number indicates there are 12 teeth in the gear for each inch of pitch diameter. Stated another way, diametral-pitch numbers specify the number of teeth in 3.1416 inches along the gear's pitch line.

Fig. 52 shows a gear with 4 inches of pitch diameter and its 3.1416 inches of pitch-circle circumference for each 1 inch of pitch diameter. In addition it illustrates that specifying the number of teeth for 1 inch of pitch diameter is in fact specifying the number of teeth in 3.1416 inches along the pitch line. The reason for this, is that for each 1 inch of pitch diameter there are pi (π) inches or 3.1416 inches of pitch-circle circumference.

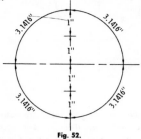

Fig. 52.

The fact that the diametral pitch number specifies the number of teeth in 3.1416 inches along the pitch line may be more easily visualized when applied to the rack. As shown in Fig. 53, the pitch line of a rack is a straight line and a measurement may be easily made along it.

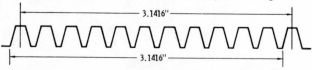

Fig. 53.

Diametral Pitch Approximation

In cases where it is necessary to determine the diametral pitch of a gear, this may be done easily without the use of precision measuring tools, templates or gauges. Measurements need not be exact because diametral-pitch numbers are usually whole numbers. Therefore, if an approximate calculation results in a value close to a whole number, that whole number is the diametral-pitch number of the gear. There are three easy methods of determining the approximate diametral pitch of a gear. A common steel rule, preferably flexible, is adequate to make the required measurements.

Method #1—Count the number of teeth in the gear, add two (2) to this number and divide by the outside diameter of the gear. For example, the gear shown in Fig. 54 has 40 teeth and its outside diameter is about $4\frac{7}{32}''$. Adding 2 to 40 gives 42, dividing 42 by $4\frac{7}{32}''$ gives an answer of $9\frac{31}{32}$. As this is approximately 10, it can be safely stated that the gear is a 10 diametral-pitch gear.

Method #2—Count the number of teeth in the gear and divide this number by the measured pitch diameter. The pitch diameter of the gear is measured from the root or bottom of a tooth space to the top of a tooth on the opposite side of the gear.

Using the same 40-tooth gear shown in Fig. 54, the pitch diameter measured from the bottom of a tooth space to the top of the opposite tooth is about $4''$. Dividing 40 by 4 gives an answer of 10. In this case the approximate whole-number pitch-diameter measurement results in a whole-number answer. This method also indicates that the gear is a 10 diametral-pitch gear.

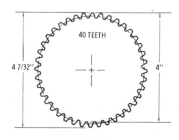

Fig. 54.

Method #3—Using a flexible scale, measure approximately 3⅛″ along the pitch line of the gear. To do this bend the scale to match the curvature of the gear and hold it about midway between the base and the top of the teeth. This will place the scale approximately on the pitch line of the gear. If the gear can be rotated, draw a pencil line on the gear to indicate the pitch line, this will aid in positioning the scale. Count the teeth in 3⅛″ to determine the diametral pitch number of the gear.

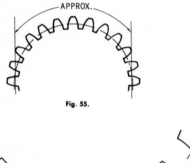

Fig. 55.

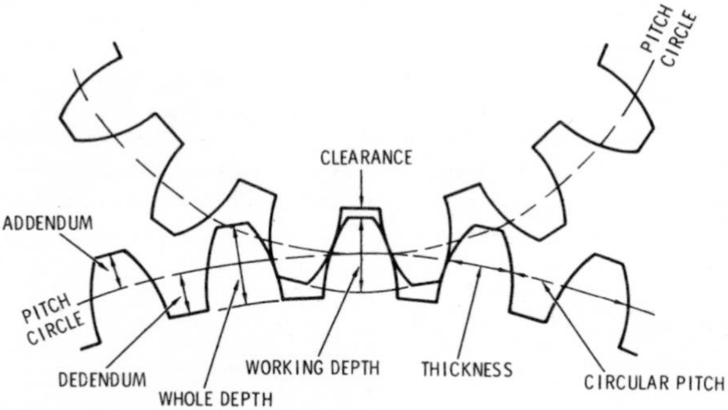

Fig. 56.

Gear Tooth Parts (Fig. 56)

Addendum—The distance the tooth projects above, or outside of the pitch line or circle.

Dedendum—The depth of a tooth space below, or inside of the pitch line or circle.

Clearance—The amount by which the dedendum of a gear tooth exceeds the addendum of a mating gear tooth.

Whole Depth—The total height of a tooth or the total depth of a tooth space.

Working Depth—The depth of tooth engagement of two mating gears. It is the sum of their addendums.

Tooth Thickness—The distance along the pitch line or circle from one side of a gear tooth to the other.

Gear Dimensions

The *outside diameter* of a spur gear is the pitch diameter plus two addendums.

The *bottom*, or *root diameter*, of a spur gear is the outside diameter minus two whole depths.

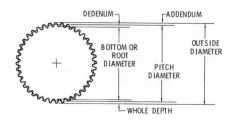

Fig. 57.

$$\text{Number of Teeth} = N$$

$$\text{Diametral Pitch} = P = \frac{N}{D}$$

$$\text{Pitch Diameter} = D = \frac{N}{P}$$

$$\text{Addendum} = A = \frac{1}{P}$$

$$\text{Whole Depth} = WD = \frac{2.2}{P} + .002''$$

$$\text{Outside Diameter} = OD = D + 2A$$

$$\text{Root Diameter} = RD = OD - 2WD$$

Backlash

Backlash in gears is the play between teeth that prevents binding. In terms of tooth dimensions, it is the amount by which the width of tooth spaces exceed the thickness of the mating gear teeth. Backlash may also be described as the distance, measured along the pitch line, that a gear will move when engaged with another gear that is fixed or unmovable.

Normally there must be some backlash present in gear drives to provide running clearance. This is necessary as binding of mating gears can result in heat generation, noise, abnormal wear, possible overload, and/or failure of the drive. A small amount of backlash is also desirable because of the dimensional variations involved in manufacturing tolerances.

Backlash is built into standard gears during manufacture by cutting the gear teeth thinner than normal by an amount equal to one half the required figure. When two gears made in this manner are run together, at standard center distance, their allowances combine to provide the full amount of backlash required.

On nonreversing drives, or drives with continuous load in one direction, the increase in backlash that results from tooth wear does not adversely affect operation. However, on reversing drives and drives where timing is critical, excessive backlash usually cannot be tolerated.

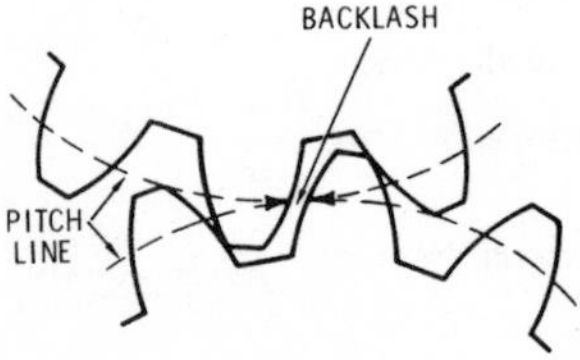

Table 9.

Fig. 58.

Pitch	Backlash
3 P	.013
4 P	.010
5 P	.008
6 P	.007
7 P	.006
8 - 9 P	.005
10 - 13 P	.004
14 - 32 P	.003
33 - 64 P	.0025

Table 9 lists the suggested backlash for a pair of gears operating at standard center distance.

COUPLINGS

The usual means of connecting two coaxial shafts so that one may transmit power to the other is by some type of power transmission coupling. Such couplings are manufactured in a great variety of types and styles, however, they may be divided into two general groupings, *rigid* (also called *solid*) couplings and *flexible* couplings.

Rigid couplings, as the name indicates, are used to connect shaft ends together rigidly. They provide a fixed union that is equivalent to a shaft extension with no movement at the joint. There are two important rules that should be followed to obtain satisfactory service from rigid couplings.

1. Use a force fit to assemble the coupling halves to the shaft ends.
2. After assembly check mating surfaces for run-out. Machine-true any surfaces that show run-out.

Flexible couplings are used to connect coaxial shafts of independent units, as they can accommodate some misalignment and end float while they transmit power and positive rotation. They are intended to overcome slight unavoidable errors in alignment, not to connect shafts on different planes or at an angle.

Independent units connected with flexible couplings, such as motors to reducers or pumps, must be aligned when installed. This alignment operation, commonly called "coupling alignment," is actually the alignment of the center lines of the coaxial shafts the coupling connects. Alignment operations are performed on the coupling because it provides convenient surfaces at the shaft ends to use for this purpose.

Shaft center lines may be out of alignment in two ways. They may be at an angle rather than in a straight line, called *angular misalignment*. Or they may be offset, called *parallel misalignment*. (Fig. 59).

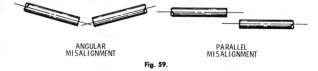

ANGULAR
MISALIGNMENT

PARALLEL
MISALIGNMENT

Fig. 59.

Misalignment may be in any plane through 360 degrees. A practical method of bringing center lines into true alignment is to check and align them in two planes at right angles. The practice commonly fol-

lowed, because it is convenient, is to align the shaft center lines in the
vertical and the horizontal planes as illustrated in Fig. 60.

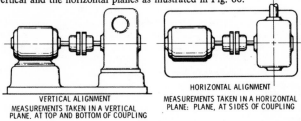

VERTICAL ALIGNMENT
MEASUREMENTS TAKEN IN A VERTICAL
PLANE, AT TOP AND BOTTOM OF COUPLING

HORIZONTAL ALIGNMENT
MEASUREMENTS TAKEN IN A HORIZONTAL
PLANE: PLANE, AT SIDES OF COUPLING

Fig. 60.

Coupling Alignment

The alignment of coaxial shafts, connected by a flexible coupling, is
commonly called *coupling alignment*. Because the alignment operations
are performed on the coupling surfaces it is extremely important that
these surfaces, and the shafts, run true. If there is any run-out of the
shaft or coupling surfaces, a proportionate error in alignment will be
assembled into the coupling. Therefore, prior to any alignment measurements and adjustments the condition of the shafts and coupling surfaces should be checked and runout eliminated if present. Another
important preparatory step is to check the footing of the units. If there
is any rocking motion the open point must be shimmed so all bearing
points rest solidly on the base plate.

Two methods of coupling alignment are widely used: "The Straightedge—Feeler Gauge Method" and "The Indicator Method." In both
methods four (4) alignment operations are performed in a specific
order. Only when performed in the correct order can adjustments be
made at each step without disturbing prior settings. The four steps in
the order of performance are:

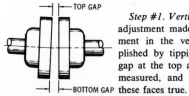

Step #1. Vertical Face Alignment—The first
adjustment made to correct angular misalignment in the vertical plane. This is accomplished by tipping the unit as required. The
gap at the top and bottom of the coupling is
measured, and adjustment is made to bring
these faces true.

Step #2. Vertical Height Alignment—This adjustment corrects parallel misalignment in the vertical plane. The unit is raised without changing its angular position. Height difference from base to center line is determined by measuring on the OD of the coupling at top and/or bottom.

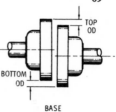

Steps #3 and #4. Horizontal Face and OD Alignment—When the units are in alignment vertically, shimming is complete. The horizontal alignment operations may then be done simultaneously. The unit is moved as required to align the faces and OD's at the sides of the coupling.

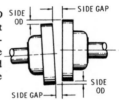

Vertical Face Alignment Correction

The initial alignment step *(Vertical Face Alignment)* can be a time consuming operation when done in an unorganized manner. Correction of angular misalignment requires tilting one of the units into correct position using shims. Selection of shim thickness is commonly done by trial and error. Much time can be saved (and accuracy gained) by using simple proportion to determine the shim thickness.

The tilt required, or the angle of change at the base, is the same as the angle of misalignment at the coupling faces. Because of this angular relationship, the shim thickness is proportional to the misalignment. For example, in Fig. 61 the misalignment at the coupling faces is 0.006″ in 5 inches, therefore each 5 inches of base must be tilted 0.006″ to correct misalignment. As the base length is twice 5 inches the shim thickness must be twice 0.006″ or 0.012″.

A simple rule for shim-thickness selection that gives accurate results and enables correction to be made in one setting is: *Shim thickness is as many times greater than the misalignment as the base length is greater than the coupling diameter.*

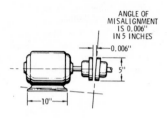

Fig. 61.

Selection Of Unit To Be Adjusted

In preparation to aligning a coupling it must be determined which
unit is to be adjusted—the "driver" or the "driven." Common practice,
and the one generally recommended is to position, level, and secure the
driven unit at the required elevation. Then, adjust the driver to align
with it. Connections to the driven unit, such as pipe connections to a
pump or output shaft connections to a reducer, should be completed
prior to proceeding with coupling alignment. The driven unit should be
set with its shaft center line slightly higher than the driver to allow for
alignment shims.

The Straightedge-Feeler Gauge Method—Practically all flexible
couplings on drives operating at average speeds will perform satisfactorily when misaligned as much as .005". Some will tolerate much greater
misalignment. Alignment well within .005" is easily and quickly attain-
able using straight edge and feeler gauge when correct methods are
followed.

Step #1. Vertical Face Alignment—Using the feeler gauge, measure
the width of gap at top and bottom between the coupling faces. Using
the difference between the two measurements determine the shim thickness required to correct alignment. (It will be as many times greater
than the misalignment as the driver base length is greater than the
coupling diameter). Shim under low end of driver to tilt into alignment
with the driven unit.

Example: Assume that measured misalignment is .160″ minus .152″ equaling .008″ misalignment in 5 inches. Base length is about 2½ times coupling diameter therefore, shim required is 2½ times .008″ or .020″. A shim .020″ thick placed under the low end

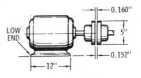

Fig. 62.

of the driver will tilt it into approximate angular alignment with the driven unit. The slight error involved is well within tolerance of flexible coupling alignment.

Step #2. Vertical Height Alignment—Using a straight edge and feeler gauge, measure the height difference between driver and driven units on the OD surfaces of the coupling. Place shims at all driver support points equal in thickness to the measured height difference.

Step #3 and #4. Horizontal Face and OD Alignment—Using a straight edge check alignment of OD's at sides of coupling. Using a feeler gauge check the gap between coupling faces at the sides of the coupling. Adjust driver as necessary to align the OD's and to set the gap equal at the sides. Do not disturb shims.

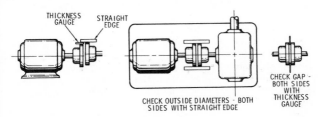

Fig. 63.

The Indicator Method (Fig. 64)

When an indicator is used as a measuring instrument to align coaxial shafts, it must be attached to the shaft or coupling of one of the units so it can be rotated. A stationary indicator will only show runout, when in contact with an object being rotated. When the indicator is rotated its tip will describe a circle which is concentric with the shaft bearings. The mating unit shaft should also be rotated as the point contacted by the indicator tip will also describe a true circle. When these two circles coincide (zero runout on the indicator) the shafts will be in alignment in the plane being measured.

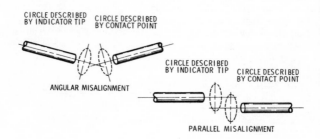

Fig. 64.

While the ideal practice is to rotate both units while indicating, this is sometimes not possible. In such cases the indicator is attached to the unit that can be rotated and the mating unit remains stationary. When this is done the surfaces of the coupling half remaining stationary must be true. Any surface errors or runout will result in a corresponding error in shaft alignment.

If both units are rotated while measuring, coupling surface runout does not affect accuracy of alignment. This is an advantage of the indicator alignment method as it is possible to accurately align shafts in spite of coupling surfaces that are not running true.

Step #1. Vertical Face Alignment—Attach indicator to shaft or coupling half of driver. Place indicator tip in contact with the face of the coupling half on the driven unit. Rotate shafts of both units together. Note indicator readings at top and bottom. Total indicator runout is a measure of the vertical angular misalignment. Place shims under driver at low end, tipping it into alignment with driven unit. The shim thickness will be as many times greater than the coupling face misalignment as the driver base length is greater than the coupling diameter.

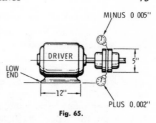

Fig. 65.

Example: Assume total indicator runout is .007″ (minus .005″ to plus .002″) as shown in Fig. 65. The driver base length is approximately 2½ times the coupling diameter (12″ to 5″). The shim thickness should be 2½ times .007″ or .017″.

Step #2. Vertical Height Alignment—Place indicator tip in contact with outside surface of driven unit coupling half. Rotate shafts of both units together. Note indicator reading at top and bottom. Height difference is one half total indicator runout. Place shims at all driver support points equal in thickness to one half total indicator runout.

Example: Assume a total indicator runout of .018″ (plus .012″ to minus .006″) as shown in Fig. 66. The required shim thickness is .009″, one half the .018″ total indicator runout.

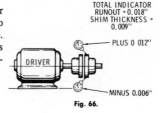

Fig. 66.

Steps #3 and #4. Horizontal Face & OD Alignment—Place indicator tip in contact with face of driven unit coupling half. Move driver as necessary to obtain zero reading on indicator. Place indicator in contact with OD surface of driven unit coupling half. Move driver as necessary to obtain zero reading on indicator. Repeat operations as necessary to obtain zero readings on both face and OD surfaces at sides of coupling. Do not disturb shims during horizontal alignment adjustments.

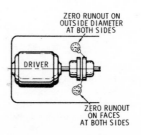

Fig. 67.

Temperature Change Compensation

To compensate for temperature difference between installation conditions and operating conditions, it may be necessary to set one unit high, or low, when aligning. For example, centrifugal pumps handling cold water, and directly connected to electric motors, require a low motor setting to compensate for expansion of the motor housing as its temperature rises. If the same units were handling liquids hotter than the motor operating temperature, it might be necessary to set the motor high. Manufacturers recommendations should be followed for initial setting when compensation for temperature change is made at cold installation.

Final alignment of equipment with appreciable operating temperature difference should be made after it has been run under actual operating conditions long enough to bring both units to operating temperatures.

SCREW THREADS

The *Unified* screw thread standards superseded the *American* standards in 1948 when an accord was signed by the standardizing bodies of Canada, the United Kingdom, and the United States. The *Unified* standards apply to the form, designation, dimensions, etc., of triangular threads in the inch system. It is substantially the same form as the *American* thread, and is mechanically interchangeable. The design form and proportions of the *Unified* thread are illustrated below.

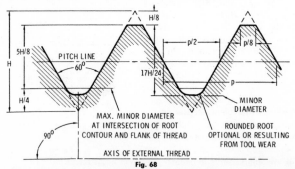

Fig. 68

Unified standards are established for various thread series; thread series being groups of diameter-pitch combinations distinguished by the number of threads per inch applied to a specific diameter.

Coarse-Thread Series—UNC—*Unified National Coarse*

Designated by the symbol *UNC* it is generally used for bolts, screws, nuts, and other general classifications.

Fine-Thread Series—UNF—*Unified National Fine*

Designated by the symbol *UNF* it is suitable for bolts, screws, nuts, etc., where a finer thread than that provided by the coarse series is required.

Extra-Fine-Thread Series—UNEF—*Unified National Ex-Fine*

Designated by the symbol *UNEF* it is used for short lengths of engagement, thin-walled tubes, ferrules, couplings, etc., where very fine pitches of threads are required.

Constant-Pitch Series—UN—*Unified National Form*

Designated by the symbol *UN,* various pitches are used on a variety of diameters. Preference is given whenever possible to the use of the 8-, 12-, and 16-thread series.

Thread Classes

The *Unified* standard establishes limits of tolerance called "classes." Classes 1A, 2A, and 3A are for external threads and classes 1B, 2B, and 3B are for internal threads. Classes 1A and 1B provide the maximum fit allowances, classes 2A and 2B provide optimum fit allowances and classes 3A and 3B provide minimum allowances.

Classes 2A and 2B are the most commonly used standards for general applications and production items such as bolts, screws, nuts, etc. Classes 3A and 3B are used when close tolerances are desired. Classes 1A and 1B are used where a liberal allowance is required to permit ready assembly, even with slightly bruised or dirty threads.

Unified Thread Designation

The standard method of designation is to specify in sequence the nominal size, number of threads per inch, thread series symbol and thread class symbol. For example a ¾ inch Unified coarse series thread for a common fastener would be designated as follows:

```
        ┌─Nominal Size
        │  ┌─Number of Threads Per Inch
        │  │  ┌Thread Series Symbol
        │  │  │      ┌Thread Class Symbol
        │  │  │      │
       ¾ -10 UNC 2A
```

Unless otherwise specified threads are right hand. A left hand thread is designated by adding the letters LH after the thread class symbol.

Screw Thread Terms

Fig. 69 illustrates the meaning of the more important screw thread terms. While "lead" is not included in the illustration, it also is an important screw thread term.

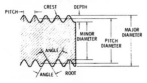

Fig. 69

Lead—The distance a screw advances axially in one turn. On a single-thread screw, the lead and pitch are the same. On a double-thread screw, the lead is twice the pitch. On a triple-thread screw, the lead is three times the pitch etc.

Translation Threads

Screw threads used to move machine parts for adjustment, setting, transmission of power, etc., are classified as "translation threads." To perform these functions a stronger form than the triangular "V" is often required. The "Square", "Acme" and the "Buttress" thread forms have the required strength and are widely used for translation thread applications.

Square Thread

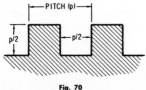

Fig. 70

The square thread is a strong and efficient thread but it is difficult to manufacture. The theoretical proportions of an external square thread are shown in Fig. 70. The mating nut must have a slightly larger thread space than the screw to allow a sliding fit. Similar clearance must also be provided on the major and minor diameters.

Table 10. Unified Standard Screw Thread Series

Sizes		BASIC MAJOR DIAMETER	THREADS PER INCH							
			Series with graded pitches			Series with constant pitches				
Prim.	Sec.		Coarse UNC	Fine UNF	Extra fine UNEF	4UN	6UN	8UN	12UN	16UN
0		0.0600	—	80	—	—	—	—	—	—
	1	0.0730	64	72	—	—	—	—	—	—
2		0.0860	56	64	—	—	—	—	—	—
	3	0.0990	48	56	—	—	—	—	—	—
4		0.1120	40	48	—	—	—	—	—	—
5		0.1250	40	44	—	—	—	—	—	—
6		0.1380	32	40	—	—	—	—	—	—
8		0.1640	32	36	—	—	—	—	—	—
10		0.1900	24	32	—	—	—	—	—	—
	12	0.2160	24	28	32	—	—	—	—	—
¼		0.2500	20	28	32	—	—	—	—	—
⁵⁄₁₆		0.3125	18	24	32	—	—	—	—	—
⅜		0.3750	16	24	32	—	—	—	—	UNC
⁷⁄₁₆		0.4375	14	20	28	—	—	—	—	16
½		0.5000	13	20	28	—	—	—	—	16
⁹⁄₁₆		0.5625	12	18	24	—	—	—	UNC	16
⅝		0.6250	11	18	24	—	—	—	12	16
	¹¹⁄₁₆	0.6875	—	—	24	—	—	—	12	16
¾		0.7500	10	16	20	—	—	—	12	UNF
	¹³⁄₁₆	0.8125	—	—	20	—	—	—	12	16
⅞		0.8750	9	14	20	—	—	—	12	16
	¹⁵⁄₁₆	0.9375	—	—	20	—	—	—	12	16
1		1.0000	8	12	20	—	—	UNC	UNF	16
	1¹⁄₁₆	1.0625	—	—	18	—	—	8	12	16
1⅛		1.1250	7	12	18	—	—	8	UNF	16
	1³⁄₁₆	1.1875	—	—	18	—	—	8	12	16
1¼		1.2500	7	12	18	—	—	8	UNF	16
	1⁵⁄₁₆	1.3125	—	—	18	—	—	8	12	16
1⅜		1.3750	6	12	18	—	UNC	8	UNF	16
	1⁷⁄₁₆	1.4375	—	—	18	—	6	8	12	16
1½		1.5000	6	12	18	—	UNC	8	UNF	16
	1⁹⁄₁₆	1.5625	—	—	18	—	6	8	12	16
1⅝		1.6250	—	—	18	—	6	8	12	16
	1¹¹⁄₁₆	1.6875	—	—	18	—	6	8	12	16
1¾		1.7500	5	—	—	—	6	8	12	16
	1¹³⁄₁₆	1.8125	—	—	—	—	6	8	12	16
1⅞		1.8750	—	—	—	—	6	8	12	16
	1¹⁵⁄₁₆	1.9375	—	—	—	—	6	8	12	16
2		2.0000	4½	—	—	—	6	8	12	16
	2⅛	2.1250	—	—	—	—	6	8	12	16
2¼		2.2500	4½	—	—	—	6	8	12	16
	2⅜	2.3750	—	—	—	—	6	8	12	16

Table 10. Unified Standard Screw Thread Series—(Continued)

Sizes		BASIC MAJOR DIAMETER	Series with graded pitches			THREADS PER INCH				
						Series with constant pitches				
Prim.	Sec.		Coarse UNC	Fine UNF	Extra fine UNEF	4UN	6UN	8UN	12UN	16UN
2½		2.5000	4	—	—	UNC	6	8	12	16
	2⅝	2.6250	—	—	—	4	6	8	12	16
2¾		2.7500	4	—	—	UNC	6	8	12	16
	2⅞	2.8750	—	—	—	4	6	8	12	16
3		3.0000	4	—	—	UNC	6	8	12	16
	3⅛	3.1250	—	—	—	4	6	8	12	16
3¼		3.2500	4	—	—	UNC	6	8	12	16
	3⅜	3.3750	—	—	—	4	6	8	12	16
3½		3.5000	4	—	—	UNC	6	8	12	16
	3⅝	3.6250	—	—	—	4	6	8	12	16
3¾		3.7500	4	—	—	UNC	6	8	12	16
	3⅞	3.8750	—	—	—	4	6	8	12	16
4		4.0000	4	—	—	UNC	6	8	12	16
	4⅛	4.1250	—	—	—	4	6	8	12	16
4¼		4.2500	—	—	—	4	6	8	12	16
	4⅜	4.3750	—	—	—	4	6	8	12	16
4½		4.5000	—	—	—	4	6	8	12	16
	4⅝	4.6250	—	—	—	4	6	8	12	16
4¾		4.7500	—	—	—	4	6	8	12	16
	4⅞	4.8750	—	—	—	4	6	8	12	16
5		5.0000	—	—	—	4	6	8	12	16
	5⅛	5.1250	—	—	—	4	6	8	12	16
5¼		5.2500	—	—	—	4	6	8	12	16
	5⅜	5.3750	—	—	—	4	6	8	12	16
5½		5.5000	—	—	—	4	6	8	12	16
	5⅝	5.6250	—	—	—	4	6	8	12	16
5¾		5.7500	—	—	—	4	6	8	12	16
	5⅞	5.8750	—	—	—	4	6	8	12	16
6		6.0000	—	—	—	4	6	8	12	16

Acme Thread

The acme thread, while not quite as strong as the square thread, is preferred because it is fairly easy to machine. The angle of an acme thread, measured in an axial plane, is 29 degrees. The basic proportions of an acme thread are shown in Fig. 71. Standards for acme screw threads establishes thread series, fit classes, allowances and tolerances, etc., similar to the standards for the unified thread form.

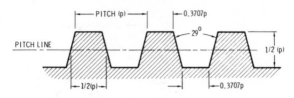

Fig. 71

Buttress Thread

The buttress thread has one side cut approximately square and the other side slanting. It is used when a thread having great strength along the thread axis in one direction only is required. Because one side is cut nearly perpendicular to the thread axis, there is practically no radial thrust when the thread is tightened. This feature makes the thread form particularly applicable where relatively thin tubu-

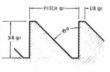

Fig. 72

lar members are screwed together. The basic thread form of a simple design of buttress thread is shown in Fig. 72. Other buttress thread forms are complex with the load side of the thread inclined from the perpendicular to facilitate machining. The angle of inclination of the American Standard buttress thread form is 7 degrees.

Metric Threads

The International Standards (ISO) and the German Standards (DIN) metric screw thread form is shown in Fig. 73. Metric threads are designated by specifying in sequence the capital letter M to signify metric, the major diameter in millimeters, and the pitch in millimeters, as shown below.

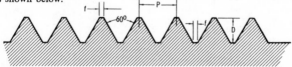

Fig. 73

SCREW THREADS

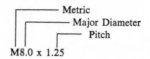

Metric
Major Diameter
Pitch

M8.0 x 1.25

INTERNATIONAL STANDARD—METRIC
THREAD DIMENSIONS AND TAP DRILL SIZES

Major Diameter m/m	Pitch m/m	Minor Diameter m/m	Pitch Diameter m/m	Tap Drill for 75% Thread m/m	Tap Drill for 75% Thread No. of Inches	Clearance Drill Size
2.0	.40	1.48	1.740	1.6	1/16	41
2.3	.40	1.78	2.040	1.9	48	36
2.6	.45	2.02	2.308	2.1	45	31
3.0	.50	2.35	2.675	2.5	40	29
3.5	.60	2.72	3.110	2.9	33	23
4.0	.70	3.09	3.545	3.3	30	16
4.5	.75	3.53	4.013	3.75	26	10
5.0	.80	3.96	4.480	4.2	19	3
5.5	.90	4.33	4.915	4.6	14	15/64"
6.0	1.00	4.70	5.350	5.0	9	1/4"
7.0	1.00	5.70	6.350	6.0	15/64"	19/64"
8.0	1.25	6.38	7.188	6.8	H	11/32"
9.0	1.25	7.38	8.188	7.8	5/16"	3/8"
10.0	1.50	8.05	9.026	8.6	R	27/64"
11.0	1.50	9.05	10.026	9.6	V	29/64"
12.0	1.75	9.73	10.863	10.5	Z	1/2"
14.0*	1.25	12.38	13.188	13.0	33/64"	9/16"
14.0	2.00	11.40	12.701	12.0	15/32"	9/16"
16.0	2.00	13.40	14.701	14.0	35/64"	21/32"
18.0*	1.50	16.05	17.026	16.5	41/64"	47/64"
18.0	2.50	14.75	16.376	15.5	39/64"	47/64"
20.0	2.50	16.75	18.376	17.5	11/16"	13/16"
22.0	2.50	18.75	20.376	19.5	49/64"	57/64"
24.0	3.00	20.10	22.051	21.0	53/64"	31/32"
27.0	3.00	23.10	25.051	24.0	15/16"	1 3/32"
30.0	3.50	25.45	27.727	26.5	1 3/64"	1 13/64"
33.0	3.50	28.45	30.727	29.5	1 11/64"	1 21/64"
36.0	4.00	30.80	33.402	32.0	1 17/64"	1 7/16"
39.0	4.00	33.80	36.402	35.0	1 3/8"	1 9/16"
42.0	4.50	36.15	39.077	37.0	1 29/64"	1 43/64"
45.0	4.50	39.15	42.077	40.0	1 37/64"	1 13/16"
48.0	5.00	41.50	44.752	43.0	1 11/16"	1 29/32"

*Special spark-plug sizes.

Thread Tapping

The cutting of internal screw threads with a hand tap is potentially a troublesome operation involving broken taps and time-consuming efforts to remove them. These troubles may be largely avoided by a better understanding of the tapping operation and the thread cutting tool. Fig. 74 illustrates a typical hand tap and gives the names of its various parts.

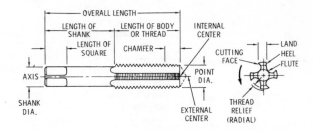

Fig. 74

The threaded body of a tap is composed of lands, which are the cutters, and flutes or channels to let the chips out and permit cutting oil to reach the cutting edges. It is chamfered at the point to allow it to enter a hole and to spread the heavy cutting operation over several rings of lands or cutting edges.

The radial relief shown in Fig. 74 refers to material having been removed from behind the cutting edges to provide clearance and reduce friction. Another form of clearance provided on commercial taps is termed "back taper." This is done by a very slight reduction in thread diameter at the shank.

The three standard styles of hand taps, *taper, plug,* and *bottoming,* have varying amounts of point "chamfer." The taper taps have the longest (8 to 10 threads) and are usually used for starting a tapped hole. The plug tap has 3 to 4 threads chamfer and is used to provide full threads more closely to the bottom of a hole than is possible with a taper tap. The bottoming tap has practically no chamfer, but when preceded by a plug tap and carefully used, can cut threads very close to the bottom of a hole.

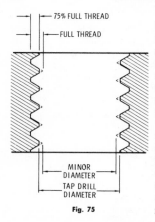

Fig. 75

The hole for the tap must be made by a *tap drill* larger than the minor diameter of the thread. The hole is made oversize to provide clearance between the wall of the hole and the roots of the tap threads. This gives chip space and allows free turning of the tap, reducing the tendency for the threads to tear.

The amount of hole oversize is such that the internal thread will be 75% of standard, as shown in Fig. 75. The 25% that is missing from the crest does not appreciably reduce its strength. Tap drill charts normally carry the notation: "Based on approximately 75% of full thread."

MECHANICAL FASTENERS

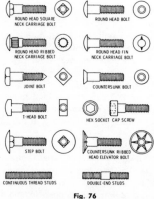

Fig. 76

Machinery and equipment is assembled and held together by a wide variety of fastening devices. Threaded fasteners are by far the most widely used style, and the bolt, screw and stud are the most common of the threaded fasteners.

The bolt is normally tightened and released by turning a mating nut. A screw differs from a bolt in that it is supposed to mate with an internal thread into which it is tightened or released by turning its head. Obviously these descriptions do not always apply, as bolts can be screwed into threaded holes and screws can be used with a nut. The stud is simply a cylindrical rod threaded on one or either ends or throughout its entire length. Some of the common bolts, screws, and stud types are shown in Fig. 76.

Threaded fasteners are furnished with either coarse threads conforming to *Unified National Coarse (UNC)* standards, or fine threads conforming to *Unified National Fine (UNF)* standards.

Coarse Threads

Fasteners with coarse threads are used for the majority of applications because they have the following advantages: They assemble easy and fast, providing a good start with little chance of cross threading; nicks and burrs from handling are not liable to affect assembly; the seizing probability in temperature applications and in joints where corrosion will form is low; They are not prone to strip when threaded into lower strength materials; The coarse thread is more easily tapped in brittle materials and materials that crumble easily.

Fine Threads

The use of fine threads may provide superior fasteners for applications where strength or other specific qualities are required. Fine threads have the following advantages: They are 10% stronger than coarse threads because of greater cross-sectional area; fine threads tap easier in very hard materials; they can be adjusted more precisely because their smaller helix angle; they may be used with thinner wall thicknesses; and they are less liable to loosening from vibration.

Washers

Most threaded fasteners are installed where vibration occurs. This motion tends to overcome the frictional force between the threads causing the fastener to back off and loosen. Washers are placed beneath the fastener head to help maintain frictional resistance to loosening.

Flat washers provide a bearing surface and spread the load over an increased holding area. Lockwashers tend to retard loosening of inadequately tightened fasteners. Theoretically, if fasteners are properly tightened, lockwashers would

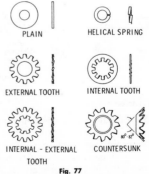

PLAIN   HELICAL SPRING

EXTERNAL TOOTH   INTERNAL TOOTH

INTERNAL - EXTERNAL   COUNTERSUNK
TOOTH

Fig. 77

not be necessary. Multiple tooth locking washers provide a greater resistance to loosening because their teeth bite into the surface against which the head or nut bears. Their teeth are twisted to slide against the surface when tightened and hold when there is a tendency to loosen. They are most effective when the mating surface under the teeth is soft.

Nuts and Pins

The nut is the mating unit used with bolt type fasteners to produce tension by rotating and advancing on the bolt threads. Nuts should be of equal grade of metal with the bolt to provide satisfactory service. Many types of nuts are available, those in common use are illustrated in Fig. 78.

Pins may be inserted through the nut and bolt after tightening to prevent the nut from turning. Some of the pins shown below are also used for other applications such as shear pins, locating and positioning parts, hinge applications, etc.

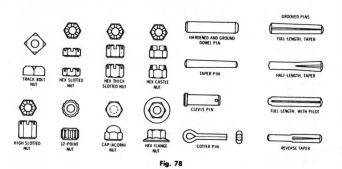

Fig. 78

Characteristics

Mechanical fasteners are manufactured in a great variety of types and sizes to suit application requirements. However, their design features, called *characteristics,* are standardized. Fasteners with almost any combination of these characteristics are commercially available. Chart 3 shows many of the characteristics that distinguish one fastener from another.

Measurements

Threaded fasteners are identified by their nominal diameter and one or more of the measurements illustrated in Fig. 79.

Chart 3. Characteristics of Standard Fasteners

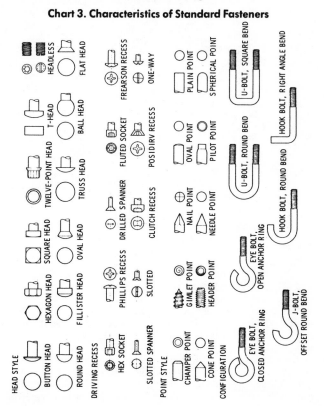

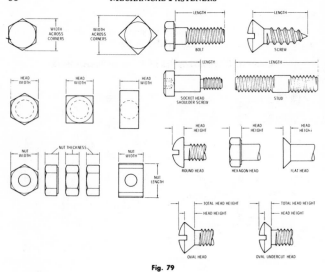

Fig. 79

Retaining Rings

Threaded fasteners are being replaced in an ever increasing number of applications by metal "retaining rings." Because these rings are installed in grooves which often can be machined simultaneously with other production processes they eliminate threading, tapping, drilling and other machining operations. In addition to reducing manufacturing costs, the rings often provide a more compact and functional design, and in some cases make possible assemblies that would otherwise be impractical.

Retaining rings in a wide variety of styles serve as shoulders for accurately locating, retaining, or locking components on shafts and in boxes and housings. Assembly and disassembly is accomplished by expanding the external rings for assembly over a shaft, or by compressing the internal ring for insertion into a box or housing.

The names, size ranges and series numbers for some of the commonly used *Waldes Truarc* retaining rings are listed in Chart 4.

Chart 4. Retainer Ring Types

INTERNAL	EXTERNAL	EXTERNAL	EXTERNAL	EXTERNAL
BASIC **N5000** For housings and bores — Size Range .250—10.0 in. 6.4—254.0 mm.	**BOWED** **5101** For shafts and pins — Size Range .188—1.500 in. 1.8—38.1 mm.	**REINFORCED** **5115** For shafts and pins — Size Range .094—1.0 in. •	**TRIANGULAR NUT** **5300** For threaded parts — Size Range 6-32 and 8-32 10-24 and 10-32 1/4-20 and 1/4-28
BOWED **N5001** For housings and bores — Size Range 6.4—44.4 mm.	**BEVELED** **5102** For shafts and pins — Size Range 1.0—10.0 in. 25.4—254.0 mm.	**BOWED E-RING** **5131** For shafts and pins — Size Range .110—1.375 in. 2.8—34.9 mm.	**KLIPRING®** **5304** For shafts and pins — Size Range .188—1.000 in. 4.8—24.5 mm.
BEVELED **N5002** For housings and bores — Size Range 1.0—10.0 in. 25.4—254.0 mm.	**CRESCENT®** **5103** For shafts and pins — Size Range .125—2.0 in. 3.2—50.8 mm.	**E-RING** **5133** For shafts and pins — Size Range .040—1.375 in. 1.0—34.9 mm.	**TRIANGULAR** **5305** For shafts and pins — Size Range .062—.438 in. •
CIRCULAR **5005** For housings and bores — Size Range .312—2.0 in. •	**CIRCULAR** **5105** For shafts and pins — Size Range .094—1.0 in. •	**PRONG-LOCK®** **5139** For shafts and pins — Size Range .092—.438 in. •	**GRIPRING®** **5555** For shafts and pins — Size Range .079—.750 in. 2.0—19.0 mm.
INVERTED **5008** For housings and bores — Size Range .750—4.0 in. 19.0—101.6 mm.	**INTERLOCKING** **5107** For shafts and pins — Size Range .469—3.375 in. 11.9—85.7 mm.	**REINFORCED E-RING** **5144** For shafts and pins — Size Range .094—.562 in. 2.4—14.3 mm.	**HIGH-STRENGTH** **5560** For shafts and pins — Size Range .101—.328 in. •
BASIC **5100** For shafts and pins — Size Range .125—10.0 in. 3.2—254.0 mm. EXTERNAL	**INVERTED** **5108** For shafts and pins — Size Range .500—4.0 in. 12.7—101.6 mm.	**HEAVY-DUTY** **5160** For shafts and pins — Size Range .394—2.0 in. 10.0—50.8 mm.	**PERMANENT SHOULDER** **5590** For shafts and pins — Size Range .250—.750 6.4—19.0 mm.

Pliers are normally used for field installation and disassembly of retaining rings. They are designed to grasp the ring securely by the lugs and expand or compress it. Most standard pliers have straight tips, however for applications where space is limited, bent tips are available.

PACKINGS AND SEALS

Stuffing Box Packings

The oldest and still one of the most widely used shaft seals is the mechanical arrangement called a *stuffing box*. It is used to control leakage along a shaft or rod. It is composed of three parts: The *packing chamber* also called the *box*—the *packing rings*—the *gland follower* also called the *stuffing gland*.

Sealing is accomplished by squeezing the packing between the throat or bottom of the box and the gland. The packing is subjected to compressive forces which cause it to flow outward to seal against the bore of the box and inward to seal against the surface of the shaft or rod.

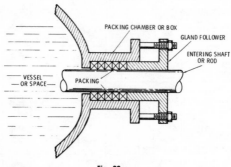

Fig. 80

Leakage along the shaft is controlled by the intimate contact of the packings on the surface of the shaft. Leakage through the packing is prevented by the lubricant contained in the packings. The packing material is called "soft" or "compression" packing and is manufactured from various forms of fibers and impregnated with binders and lubricant. The impregnated lubricant comprises about 30% of the total packing volume.

As the gland is tightened the packing is compressed and wears, therefore it must have the ability to deform in order to seal. It must also have a certain ruggedness of construction so that it may be readily cut into rings and assembled into the stuffing box without serious breakage or deformation.

Packings require frequent adjustment to compensate for the wear and loss of volume that occurs continuously while they are subjected to operating conditions. A fundamental rule for satisfactory operation of a stuffing box is: *There must be controlled leakage*. This is necessary because in operation a stuffing box is a form of braking mechanism and generates heat. The frictional heat is held to a minimum by the use of smooth polished shaft surfaces and a continuous supply of packing lubricant to the shaft-packing interface. The purpose of leakage is to assist in lubrication and to carry off the generated heat. Maintaining packing pressures at the lowest possible level helps to keep heat generation to a minimum.

Stuffing Box Arrangement

The function of the multiple rings of packing in a stuffing box is to break down the pressure of the fluid being sealed so that it approaches zero pressure, or atmosphere, at the follower end of the box. Theoretically a single ring of packing, properly installed, will seal. In practice the bottom ring in a properly installed set of packings does the major part of the sealing job. Because the bottom ring is the farthest from the follower, it has the least pressure exerted on it. Therefore, to perform its important function of a major pressure reduction, it is extremely important that it be properly installed.

Packing Installation

While the packing of a stuffing box appears to be a relatively simple operation, it is often done improperly. It is generally a hot, dirty, uncomfortable job that is completed in the shortest possible time with the least possible effort. Short packing life and damage to shaft surfaces can usually be traced to improper practices that result, rather than deficiencies in material and equipment.

For example, a common improper practice is to lay out packing material on a flat surface and cut it to measured lengths with square ends as shown in Fig. 81.

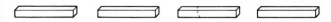

Fig. 81

When lengths of packing with square ends are formed into rings they have a pie shaped void at the ends as illustrated at right. The thicker the packing in relation to the shaft diameter the more pronounced the void will be, as the outside circumference of a ring is greater than the inside circumference. Such voids cause unequal compression and distortion of the packing. Overtightening is then required to accomplish sealing.

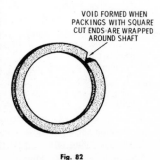

VOID FORMED WHEN PACKINGS WITH SQUARE CUT ENDS ARE WRAPPED AROUND SHAFT

Fig. 82

Correct practice is to cut the packing into rings while it is wrapped around a shaft or mandrel. A square cut end to form a plane butt joint is as satisfactory as step, angle, or scive joints.

Packing Installation

The manner of handling and installing compression packing has a greater influence on its service life than any other factor. Therefore, maximum packing life can only be realized when correct packing practices are followed. The basic steps in correct packing installation are as follows:

1. Remove all old packing and thoroughly clean the box.
2. Cut packing rings on shaft or mandrel as shown below. Keep the job simple and cut rings with plane butt joints.

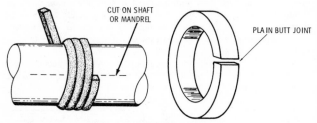

CUT ON SHAFT OR MANDREL

PLAIN BUTT JOINT

Fig. 83

3. Form the first packing gently around the shaft and enter the ends first into the box. The installation of this first packing ring is the most critical step in packing installation. The first ring should be gently pushed forward keeping the packing square with the shaft as it is being seated. It must be seated firmly against the bottom of the box with the butt ends together before additional packing rings are installed. NEVER put a few rings into the box and try to seat them with the follower. The outside rings will be damaged and the bottom rings will not be properly seated.
4. Insert additional rings individually, tamping each one firmly into position against the preceding ring. Stagger the joints to provide proper sealing and support.
5. Install the gland follower.

6. Tighten follower snugly while rotating shaft by hand. When done in this manner it is immediately apparent if the follower becomes jammed due to cocking or if the packing is overtightened. Slack off and leave finger tight.

7. Open valves to allow fluid to enter equipment. Start equipment; fluid should leak from the stuffing box. If leakage is excessive, take up slightly on the gland follower. *Do not eliminate leakage entirely;* slight leakage is required for satisfactory service. During

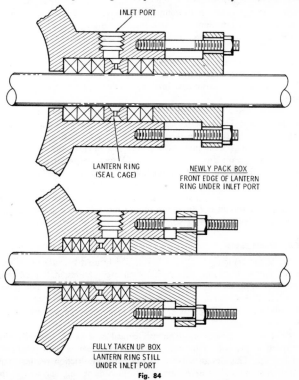

INLET PORT

LANTERN RING
(SEAL CAGE)

NEWLY PACK BOX
FRONT EDGE OF LANTERN
RING UNDER INLET PORT

FULLY TAKEN UP BOX
LANTERN RING STILL
UNDER INLET PORT

Fig. 84

the first few hours of run-in operation the equipment should be
checked periodically as additional adjustment may be required.

Stuffing Box Lantern Rings

Many stuffing box assemblies include a *lantern ring* or *seal cage*. Use
of this device allows the introduction of additional lubricants or fluids
directly to the interface of packing and shaft.

A common practice with pumps having a suction pressure below
atmospheric pressure, is to connect the pump discharge to the lantern
ring. The fluid introduced through the lantern ring acts as both a seal
to prevent air from being drawn into the pump, and as a lubricant for
the packing and shaft.

Lantern rings are also commonly used in pumps handling slurries.
In this case clear liquid from an external source, at a higher pressure
than the slurry, is introduced into the stuffing box through the lantern
ring.

Stuffing boxes incorporating lantern rings require special attention
when packing. The lantern ring must be positioned between the packing rings so its front edge is in line with the inlet port at installation.
As the packings wear and the follower is tightened, the lantern ring
will move forward under the inlet port. When the packings have been
fully compressed the lantern ring should still be in a position open to
flow from the inlet port.

While lantern rings may on occasion be troublesome to the mechanic, they should not be removed and discarded as they are an important part of the stuffing box assembly.

Mechanical Seals

While the "stuffing box" seal is still widely used because of its sim-
plicity and ability to operate under adverse conditions, it is used principally on applications where continuous slight leakage is not objec-
tionable. The *mechanical seal* operates with practically no leakage and
is replacing the stuffing box seal in an ever-increasing number of applications.

Principle of Operation—The mechanical seal is an *end-face* type
designed to provide rotary seal faces that can operate with practically
no leakage. This design uses two replaceable antifriction mating rings
—one rotating, the other stationary—to provide sealing surfaces at the

point of relative motion. These rings are statically sealed, one to the shaft, the other to the stationary housing. The mechanical seal therefore is made up of three (3) individual seals, two are static having no relative movement, the third is the rotary or dynamic seal at the end faces of the mating rings.

The mechanical seal also incorporates some form of self-contained force to hold the mating faces together. This force is usually provided by a spring-loading apparatus such as a single coil spring, multiple springs, or wave springs which are thin spring washers into which waves have been formed. The basic design and operating principle of the "inside" mechanical seal is diagrammatically illustrated in Fig. 85.

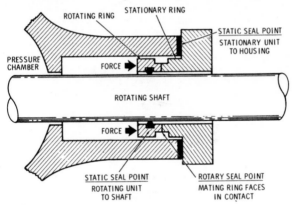

Fig. 85

Mechanical-Seal Types

While there are many design variations and numerous adaptations, there are only three basic types of mechanical seals. These are the *inside, outside,* and *double* mechanical seals.

Inside Seal—The principle of the "inside" type is shown in Fig. 85. Its rotating unit is inside the chamber or box, thus its name. Because the fluid pressure inside the box acts on the parts and adds to the force holding the faces together, the total force on the faces of an inside seal will increase as the pressure of the fluid increases. If the force built into the seal plus the hydraulic force are high enough to squeeze out the lubricating film between the mating faces, the seal will fail.

While the mechanical seal is commonly considered to be a positive seal with no leakage, this is not true. Its successful operation requires that a lubricating film be present between the mating faces. For such a film to be present there must be a very slight leakage of the fluid across the faces. While this leakage may be so slight that it is hardly visible, if it does not occur the seal will fail. This is the reason a mechanical seal must never run dry. The stuffing box must be completely filled and the seal submerged in fluid before the equipment is started, and always while operating.

Outside Seal—This style of seal is also named for the location of the rotating unit, in this case outside of the stuffing box as shown in Fig. 86. Because all the rotating parts are removed from the liquid being handled it is superior for applications where corrosive or abrasive materials are present. Because the hydraulic pressure of the fluid is imposed on the sealing faces, tending to open them by overcoming the self-contained force, it is limited to moderate pressures.

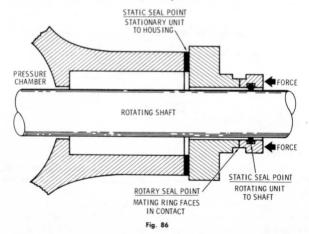

Fig. 86

Double Seal—A "double seal" is basically an arrangement of two single inside seals placed back to back inside a stuffing box. The double seal provides a high degree of safety when handling hazardous liquids. This is accomplished by circulating a non-hazardous liquid inside the box at higher pressure than the material being sealed. Any leakage therefore will be the non-hazardous lubricating fluid inward, rather than the hazardous material that is being sealed leaking outward. The basic principle of the "double mechanical seal" is illustrated in Fig. 87.

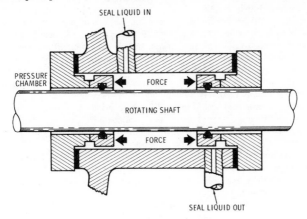

Fig. 87

Lubrication of Seal Faces—The major advantage of the mechanical seal is its low leakage rate. This is so low there is virtually no visible leakage. To operate satisfactorily in this manner requires that the film of lubricant between the seal faces be extremely thin and uninterrupted. Maintenance of this extremely thin film is made possible by machining the seal faces to very high tolerances in respect to flatness and surface finish. To protect this high quality precise surface finish requires that seal parts be carefully handled and protected. Mating faces should never be placed in contact without lubrication.

Operation of the mechanical seal depends on this thin film of lubricant furnished by leakage. If there is insufficient leakage to provide the lubricating film the faces will quickly overheat and fail. Liquid must always be present during operation as running dry for a matter of seconds can destroy the seal faces.

Mechanical Seal Installation

No one method or procedure for installation of mechanical seals can be outlined because of the variety of styles and designs. In cases where the seal is automatically located in correct position by the shape and dimensions of the parts, installation is relatively simple and straightforward. In many cases however, the location of the seal parts must be determined at installation. In such cases the location of these parts is critical, as their location determines the amount of force that will be applied to the seal faces. This force is a major factor in seal performance as excessive face pressure results in early seal failure. Parts must be located to apply sufficient force to hold the mating rings together without exerting excessive face pressure.

The procedure or method of location and installation of outside seals is usually relatively obvious and easily accomplished. The inside style of mechanical seal is more difficult to install as some parts must be located and attached while the equipment is disassembled. The location of these parts must be such that the proper force will be applied to the seal faces when the assembly is complete.

Seal designs and styles vary with manufacturers, however, the same basic principles apply to all when locating and installing inside mechanical seals. The following general procedure is applicable to most styles in common use.

Step #1—Determine the compressed length of the seal component incorporating the force mechanism. This is its overall length when it is in operating position (springs properly compressed). Two widely used seal designs are shown in Fig. 88, one incorporates multiple springs the other a single helical spring. In either case the spring or springs must be compressed the amount recommended by the seal manufacturer before the measurement to determine compressed length is made.

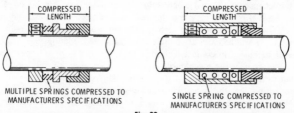

MULTIPLE SPRINGS COMPRESSED TO MANUFACTURERS SPECIFICATIONS

SINGLE SPRING COMPRESSED TO MANUFACTURERS SPECIFICATIONS

Fig. 88

This is vitally important since the force exerted on the seal faces is controlled by the amount the springs are compressed.

Manufacturers practices vary in the method of determining correct spring compression. In some cases the springs should be compressed to obtain a specific gap or space between sections of the seal assembly. In other cases it is recommended that spring or springs be compressed until alignment of lines or marks are accomplished. In any case, consult manufacturers instructions and be sure the method used to determine compressed length is correct for the make and model of seal being installed.

Step #2—Determine the insert projection of the mating seal ring. This is the distance the seal face will project into the stuffing box when it is assembled into position. Care must be exercised to be sure the static seal gasket is in position when this measurement is made. Obviously the amount of projection can be varied by varying the thickness of the gasket. See Fig. 89.

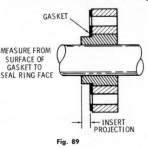

Fig. 89

Step #3—Determine the "location dimension." This is done by simply adding the "compression-length" dimension found in *Step #1* to the "insert-projection" dimension found in *Step #2*.

Step #4—"Witness-mark" the shaft in line with the face of the stuffing box. A good practice is to blue the shaft surface in the area where the mark is to be made. A flat piece of hardened steel such as a tool bit, ground on one side only to a sharp edge makes an excellent marking tool. The marker should be held flat against the face of the box and the shaft rotated in contact with it. This will provide a sharp clear witness-mark line that is exactly in line with the face of the box.

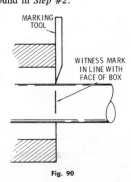

Fig. 90

Step #5—At this point the equipment must be disassembled in a manner to expose the area of the shaft where the rotary unit of the seal is to be installed. The amount and method of disassembly will vary with the design of the equipment. In some cases it may be necessary to completely remove the shaft from the equipment. In other cases, as with the back pull-out design pumps, it is only necessary to remove the back cover which contains the stuffing box chamber to expose the required area of the shaft.

Step #6—With the shaft either removed or exposed, blue the area where the back face of the rotary unit will be located. From the "witness mark," which was placed on the shaft in *Step #4*, measure the "location-dimension" distance and place a second mark on the shaft. This is called the "location mark" as it marks the point at which the back face of the rotary unit is to be located. The "location dimension" is the sum of the "compressed length" plus the "insert projection." See Fig. 93.

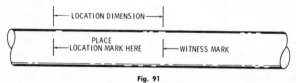

Fig. 91

Step #7—Assemble the rotary unit on the shaft with its back face on the location mark. Fasten the unit securely to the shaft at this location. Some seal designs allow separation of the rotary unit components. In such cases the back collar may be installed at this time and the other rotary unit parts later. Illustrated in Fig. 92 is a single-spring type rotary unit assembled on the shaft with its back face on the location

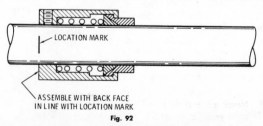

Fig. 92

mark. The spring is extended *and* will be compressed by tightening the stationary ring at assembly.

Step #8—Reassemble the equipment with the rotary unit on the shaft inside the stuffing box chamber. Complete the seal assembly, BEING SURE THE SEAL FACES ARE LUBRICATED. The sequence of parts assembly depends on the type and design of the seal and the equipment. Illustrated below is a completely installed inside seal of the multiple spring design.

The final assembly operation will be the tightening of the gland follower bolts or nuts. When this is done the lubricated seal faces should be brought into contact very carefully. When the faces initially contact there should be a space between the face of the box and the follower gland gasket. This space should be the same amount as the springs were compressed in Step #1 when the compressed length of the rotary unit was determined. This should be very carefully observed as it is a positive final check on correct location of the rotary unit of an inside mechanical seal.

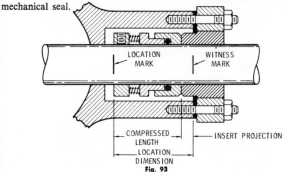

LOCATION MARK   WITNESS MARK

COMPRESSED LENGTH ── INSERT PROJECTION
LOCATION DIMENSION

Fig. 93

Installation Precautions

1. Check shaft with indicator for runout and end play. Maximum T.I.R. allowable .005 of an inch.
2. All parts must be clean and free of sharp edges and burrs.
3. All parts must fit properly without binding.
4. Inspect seal faces carefully. No nicks or scratches.
5. Never allow faces to make dry contact. Lubricate them with a good grade of oil or with the liquid to be sealed.

6. Protect all static seals such as "O" rings, "V" rings, "V" cups, wedges, etc. from damage on sharp edges during assembly.
7. Before operating, *be sure proper valves are open and seal is submerged in liquid*. If necessary vent box to expel air and allow liquid to surround seal rings.

"O" Rings

The *O-Ring* is a squeeze-type packing made from synthetic rubber compounds. It is manufactured in several shapes, the most common being the circular cross section from which it derives its name. The principle of operation of the "O" ring can be described as controlled deformation. A slight deformation of the cross-section called a "mechanical squeeze," illustrated in Fig. 94A, deforms the ring and places the material in compression. The deformation squeeze flattens the ring into intimate contact with the confining surfaces, and the internal force squeezed into the material maintains this intimate contact.

Additional deformation results from the pressure the confined fluid exerts on the surface of the material. This in turn increases the contact area and the contact pressure as shown in Fig. 94B.

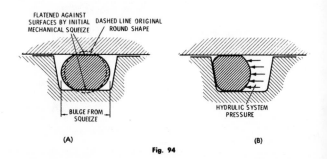

(A)   (B)

Fig. 94

The initial mechanical squeeze of the O-ring at assembly should be equal to approximately 10% of its cross-sectional dimension. The general purpose industrial O-ring is made to dimensions that in effect build the initial squeeze into the product. This is done by manufacturing the O-ring to a cross-sectional dimension 10% greater than its nominal size. Listed below are the "nominal" and "actual" cross-section diameter dimensions of O-rings in general use.

Nominal—	$\frac{1}{32}$	$\frac{3}{64}$	$\frac{1}{16}$	$\frac{3}{32}$	$\frac{1}{8}$	$\frac{3}{16}$	$\frac{1}{4}$
Actual —	.040	.050	.070	.103	.139	.210	.275

Because the cross-section dimension of an O-ring is 10% oversize, the outside and inside dimensions of the ring must be proportionately larger and smaller. For example, a 2 x 2⅜ x ³⁄₁₆ nominal size O-ring has actual dimensions of 1.975 x 2.395 x .210. Fig. 95 illustrates the relationship of such an O-ring to its groove.

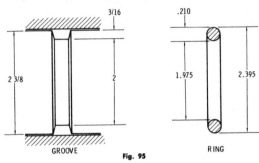

GROOVE   **Fig. 95**   RING

"O" Ring Dash Number System

Many O-ring standards have been established by various agencies, societies, industrial groups, etc. The dimensions used in all of these are essentially the same, however, the numbering systems are not. In wide use by manufacturers of O-rings is what is called the *Uniform Dash Number* system. In this system the numbers from 001 to 475 are used to identify the specific dimensions of O-rings.

Within the system of numbers the O-rings are grouped according to cross-section diameter as follows:

Dash Nos. —004 to —050 for $\frac{1}{16}$" Diameter
Dash Nos. —110 to —170 for $\frac{3}{23}$" Diameter
Dash Nos. —210 to —284 for $\frac{1}{8}$" Diameter
Dash Nos. —325 to —395 for $\frac{3}{16}$" Diameter
Dash Nos. —425 to —475 for $\frac{1}{4}$" Diameter

Using a Uniform Dash Number table one can select an O-ring of correct size from the nominal dimensions. Simply select the dash number that corresponds to the nominal dimensions of the installation. Table 11 shows the dash numbers −270 to −329. Note that −281 is the largest O-ring having a ⅛″ cross section and −325 is the smallest O-ring having a 3/16″ cross section.

Table 11

Dash No.	O-Ring Size								
	Actual						Nominal Size		
		Tolerance							
	I.D.	Table I ±	Table II ±	I.D.	±	I.D.	O.D.	W.	
−270	8.984	.030	.050	.139	.004	9	9¼	⅛	
−271	9.234	.030	.055	.139	.004	9¼	9½	⅛	
−272	9.484	.030	.055	.139	.004	9½	9¾	⅛	
−273	9.734	.030	.055	.139	.004	9¾	10	⅛	
−274	9.984	.030	.055	.139	.004	10	10¼	⅛	
−275	10.484	.030	.055	.139	.004	10½	10¾	⅛	
−276	10.984	.030	.065	.139	.004	11	11¼	⅛	
−277	11.484	.030	.065	.139	.004	11½	11¾	⅛	
−278	11.984	.030	.065	.139	.004	12	12¼	⅛	
−279	12.984	.030	.065	.139	.004	13	13¼	⅛	
−280	13.984	.030	.065	.139	.004	14	14¼	⅛	
−281	14.984	.030	.065	.139	.004	15	15¼	⅛	
───	15.955	.045	.075	.139	.004	16	16¼	⅛	
───	16.955	.045	.080	.139	.004	17	17¼	⅛	
───	17.955	.045	.085	.139	.004	18	18¼	⅛	
−325	1.475	.010	.015	.210	.005	1½	1⅞	3/16	
−326	1.600	.010	.015	.210	.005	1⅝	2	3/16	
−327	1.725	.010	.015	.210	.005	1¾	2⅛	3/16	
−328	1.850	.010	.015	.210	.005	1⅞	2¼	3/16	
−329	1.975	.010	.018	.210	.005	2	2⅜	3/16	

Formed and Molded Packings

The principle of operation of formed and molded packings is quite different from compression packings in that no compression force is required to operate them. The pressure of the fluid being sealed provides the force which seats the packings against the mating surfaces. They are therefore often classed as automatic or hydraulic packings. Packings molded or formed in the shape of a "Cup"—"Flange"—"U Shape" or—"V Shape" are classed as "lip" type packings.

Lip type packings are usually produced with lips slightly flared to provide automatic pre-load at installation. The fluid being sealed then acts against the lips exerting the force that presses them against the mating surface. Lip type packings are used almost exclusively for sealing during reciprocating motion. They must be installed in a manner that will allow the lips freedom to respond to the fluid forces.

Packing Materials

Leather—One of the oldest packing materials and still the most satisfactory for rough and difficult applications. Has high tensile strength and resistance to extrusion. Absorbs fluids, therefore tends to be self-lubricating.

Fabricated—Made by molding woven duck, asbestos cloth, and synthetic rubber. Fabric reinforcement gives strength to withstand high pressures. Also resistant to acids, alkalies and high temperature. Tends to wipe drier than leather packings although the fabric will absorb some fluid and has slight lubricating action.

Homogenous—Compounded from a wide variety of synthetic rubbers. Low strength but high resistance to acids, alkalies and high temperature. Requires a fine surface finish, close clearances, and clean operating conditions. Nonabsorbent, has no self-lubricating qualities and usually wipes contact surfaces quite dry.

Plastic—Molded from various kinds of plastics for special applications. Inert to most chemicals and solvents but has little elasticity or flexibility. Some types have a slippery feel and resist adhesion to metal but their friction under pressure is high.

Cup Packings—One of the most widely used styles of packing, simple to install and highly satisfactory for plunger end applications. Inside follower plate must not be overtightened as the bottom of the packing will be crushed and cut through. The heel or shoulder is the point of greatest wear and usually the failure point. Clearances should be held to a minimum and the lips protected from bumping. See Fig. 96.

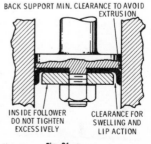

BACK SUPPORT MIN. CLEARANCE TO AVOID EXTRUSION

INSIDE FOLLOWER DO NOT TIGHTEN EXCESSIVELY

CLEARANCE FOR SWELLING AND LIP ACTION

Fig. 96

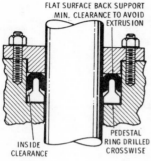

FLAT SURFACE BACK SUPPORT
MIN. CLEARANCE TO AVOID
EXTRUSION

INSIDE
CLEARANCE

PEDESTAL
RING DRILLED
CROSSWISE

Fig. 97

"U" Packings—A balanced packing as sealing occurs on both the inside and outside diameter surfaces. To support the lips the recess of the U is filled with flax, hemp, rubber, fiber, etc. Metal rings called "pedestal" rings are also used for lip support. Fillers, if nonporous, must be provided with pressure equalization openings (holes) to allow equalization of pressure on all inside surfaces. Clearance between pedestal rings and inside packing wall is necessary to accommodate swelling and allow the lips freedom to respond to the actuating fluid. See Fig. 97.

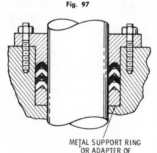

METAL SUPPORT RING
OR ADAPTER OF
LEATHER OR RUBBER

Fig. 98

"V" Packings—Installed in sets, each set consisting of a number of V-rings and a male and female adapter. Have a small cross section and are suitable for both high and low pressures. Operate as automatic packings but have the advantage of permitting taking up on the gland ring when excessive wear develops. See Fig. 98.

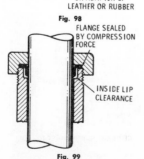

FLANGE SEALED
BY COMPRESSION
FORCE

INSIDE LIP
CLEARANCE

Fig. 99

Flange Packings—Seals on the inside diameter only, generally restricted to low pressures. Base sealed by an outside compressing force, usually a gland arrangement. Lip action same as cup packing and must be given same consideration for clearance, etc.

Packing Installation

The principle of operation of all lip-style packings is the same regardless of type or material. They must be installed in a manner that will allow them to expand and contract freely. They should not be placed under high mechanical pressure as this transforms them to compression packings. Overtightening a lip-type packing improperly preloads it. While slight preload is needed for a tight fit and sealing at low pressure, it should occur automatically as a function of the size and shape of the packing.

Lip-type packings installed for one-directional sealing in glands or on pistons should be installed with the inside of the packing exposed to the actuating fluid. Proper installation of packings in this style application is obvious if the principle of the lip-type automatic packing is understood. Double acting applications require greater care and attention at installation to ensure that assembly conforms to operating principles. When mounted with the insides of the packing facing together, they are referred to as *face-to-face* mounted. When mounted with the bottoms toward one another and the insides facing away, they are referred to as *back-to-back* mounted.

The ideal arrangement for double acting packing assembly is back-to-back with solid shoulders for back support. Each packing is fully supported and no trapping of pressure between packings can occur. Because such an arrangement requires that packings be installed from both sides of the plunger head or end, it frequently is not practical. In such cases face-to-face mounting is used. While this allows packing

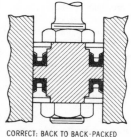

CORRECT: BACK TO BACK-PACKED
FROM TWO SIDES

Fig. 100

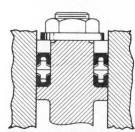

FACE TO FACE COMPROMISE TO ALLOW
PACKING FROM ONE SIDE

Fig. 101

from one end, it also introduces an undesirable condition. The actuating fluid must pass the first packing and open the second packing, which is facing toward the fluid. When the second packing is expanded by the fluid, pressure tends to back up into the first packing, expanding it and locking pressure between the two packings.

Bearings

A bearing, in mechanical terms, is a support for a revolving shaft. In some cases this may also include an assembly of several components that hold or secure the member that supports the shaft.

Bearing Nomenclature

Journal—That part of a shaft, axle, spindle, etc., which is supported by and revolves in a bearing.

Axis—The straight line (imaginary) passing through a shaft on which the shaft revolves or may be supposed to revolve.

Radial—Extending from a point or center in the manner of rays (as the spokes of a wheel are radial).

Thrust—Pressure of one part against another part, or force exerted end-wise or axially through a shaft.

Friction—Resistance to motion between two surfaces in contact.

Sliding Motion—Two parallel surfaces moving in relation to each other (Plain Bearings).

Rolling Motion—Round object rolling on mating surface with theoretically no sliding motion (antifriction bearing).

There are two broad classifications of bearings—*plain* and *antifriction*. The *plain* bearing operates on the principle of sliding motion, there being surface contact and relative movement between shaft and bearing surfaces. The antifriction bearing operates on the principle of rolling motion there being a series of rollers or balls interposed between the shaft and the supporting member.

Plain Bearings

Plain bearings are simple in design and construction, operate efficiently and are capable of supporting extremely heavy loads. During operation they develop an oil film between the journal and bearing surfaces that overcomes the friction of sliding motion. There are times however when this film is not present, starting and stopping, during shock loading or at misalignment etc. For this reason the plain bearing is made of a material softer than the shaft material and one that has low frictional qualities. The most widely used bearing materials are bronze, babbit, cast iron and plastics.

Another factor which has a major influence on plain bearing life is the surface finish of both the journal and the bearing. The rougher the surface, the thicker the film required to separate them. The high degree of surface smoothness which enables the plain bearing to operate with a very thin oil film is often achieved by a break-in or run-in period. During this period a wearing down or flattening of the peaks takes place which greatly reduces the maximum variations and smooths the surfaces.

The details of design and construction of plain bearings varies widely. They may be complete self contained units, or the bearing may be built into, and part of, a larger machine assembly. All however are variations of the following basic plain bearing types.

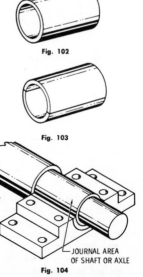

Solid Bearings—The sleeve bearing or bushing is the most common of all plain bearings. It provides the bearing surfaces for the shaft journal and is usually press-fitted into a supporting member.

Fig. 102

Split Bearings—A variation of the solid bearing, the split bearing is divided into two pieces to allow easy shaft assembly and removal.

Fig. 103

Journal Bearings — The plain bearing unit, used for support of radial loads, is commonly referred to as a "journal bearing." It takes its name from the portion of the shaft or axle that operates within the bearing.

JOURNAL AREA
OF SHAFT OR AXLE

Fig. 104

Fig. 105

Part Bearings—The part or half bearing encircling only part of the journal is used when the principal load presses in the direction of the bearing. Its advantages are low material cost and ease of replacement.

Thrust Bearings—The thrust bearing supports axial loads and/or restrains endwise movement. One widely used style is the simple annular ring or washer. Two or more such rings of selected low-frictional materials, either hard or soft, are often combined. It is also common practice to support thrust loads on the end surface of journal bearings. If the area is too small for the applied load a flange may be provided. The shaft surface which mates the thrust bearing surface and supports the axial load is usually provided by a shoulder.

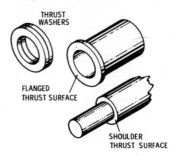

Fig. 106

Lubrication

When bearing surfaces are separated with an oil film the condition is described as *thick-film* lubrication. This film is a result of the hydraulic pressure that is generated by the rotation of the shaft. During rotation the oil is drawn into the clearance space between the journal and the shaft. The shaft is lifted on the film and separated from contact with the bearing. As the speed is increased, higher pressure develops and the shaft takes an eccentric position.

Clearance—To allow the formation of an oil film in a plain bearing there must be clearance between the journal and the bearing. This clearance varies with the size of the shaft and the bearing material, the load carried and the accuracy of shaft position desired. In industrial design of rotating machinery a diametral clearance of .001″ per inch of shaft diameter is often used. This is a general base figure and requires adjustment for high speeds or heavy loading.

Lubrication Holes and Grooves—The simplest plain bearings have a hole in the top through which the lubricant travels to the journal. For longer bearings a groove or combination of grooves extending in either direction from the

OIL GROOVE

Fig. 107

oil hole, as shown in Fig. 107, will distribute the oil.

The oil inlet hole is located in the center of the bearing in the low-pressure region. It is important that the grooving be confined to the unloaded portion of the bearing surface. If grooving extends into the loaded or pressure region, the oil film will be disrupted as the grooves will act as pressure relief passages.

Bearing Failures

Determination of the exact cause of a bearing failure is often a difficult matter. Usually the trouble lies in one or more of the following areas:

Unsuitable Materials—Under normal conditions babbit and bronze bearings may be used with soft steel journals. Harder bearing materials require a harder shaft surface.

Incorrect Grooving—If grooves are incorrectly located interfering with oil-film formation, poor lubrication and bearing failure results.

Unsuitable Finish—The smoother the finish the closer the surfaces may approach without metallic contact.

Insufficient Clearance—There must be sufficient clearance between journal and bearing to allow oil-film formation.

Operating Conditions—Speedup and overloading are the two most frequent causes of bearing failure.

Oil Contamination—Foreign material in the lubricant causes scoring and galling of the bearing surfaces.

Antifriction Bearings

Antifriction bearings are of two general types, *Ball bearings* and *roller bearings*. They operate on the principle of rolling motion, using either balls or rollers between the rotating and stationary surfaces. Because of this, friction is reduced to a fraction of that in plain bearings, thus their name "antifriction."

Ball bearings may be divided into three main groups according to function: radial, thrust, and angular contact.

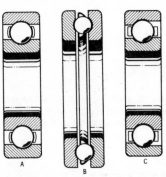

The radial bearing shown in Fig. 108A, is designed primarily to carry a load in a direction perpendicular to the axis of rotation.

The thrust bearing shown in Fig. 108B can carry thrust loads only. That is a force parallel to the axis of rotation tending to cause endwise movement of the shaft.

The angular contact bearing shown in Fig. 108C, can support combined radial and thrust loads in one direction. Angular contact bearings must be installed in pairs to support thrust loads in both directions.

Fig. 108

Roller bearings are also classified by their ability to support radial, thrust, and combination loads. In addition they are further divided into styles according to the shape of their rollers. Widely used roller styles are: *Cylindrical,* also known as *straight; taper; spherical;* and *needle.* Examples of bearings using these roller styles are shown in Fig. 109.

Combination load-supporting roller bearings are not called angular contact bearings as they are quite different in design. The taper roller bearing for example is a combination load-carrying bearing by virtue of the shape of its rollers.

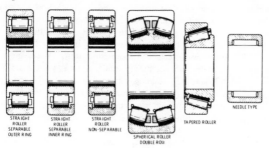

STRAIGHT ROLLER SEPARABLE OUTER RING   STRAIGHT ROLLER SEPARABLE INNER RING   STRAIGHT ROLLER NON-SEPARABLE   SPHERICAL ROLLER DOUBLE ROW   TAPERED ROLLER   NEEDLE TYPE

Fig. 109

Self-Aligning Bearings

The *self-aligning* bearing is a specialized style of antifriction bearing, which as the name indicates, has the capability of angular self-alignment. This is usually accomplished by the use of a spherical raceway inside the outer ring of the bearing. The inner ring and the rolling elements rotate at right angles to the shaft center line in a fixed raceway. The position of the outer ring, because of its spherical raceway, may be misaligned within the limits of its width and still provide a true path for the rolling elements to follow. Bearings of this style are called *internal* self-aligning.

The *spherical double-row* roller bearing in Fig. 110A is a common self-aligning style of roller bearing. Another widely used style is the *self-aligning* ball bearing at (B). The angular movement this bearing allows is possible because the two rows of balls are rolling on the spherical inner surface of the outer ring. Another self-aligning ball bearing (single row) is shown at (C).

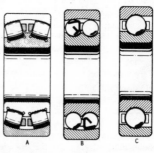

A   B   C

Fig. 110

It incorporates an additional outside ring with a spherical inner surface. The outside of the regular outer ring is made spherical to match the extra ring. This style of construction is used for single row self-aligning ball bearings. It is called an *external style* self-aligning ball bearing.

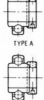

TYPE A

TYPE B

Fig. 111

Another specialized style of bearing is the *wide inner ring* bearing. It is used primarily in pillow blocks, flange units, etc., termed transmission units. Wide inner ring bearings are commonly made in two types, rigid (Type A), and self-aligning (Type E). The rigid type has a straight cylindrical outside surface, the self-aligning a spherical outside surface. They are made to millimeter outside dimensions and inch bore dimensions.

They are an assembly of standard metric size ball bearing outer rings with special wide inner rings. The bore of the inner ring is made to inch dimensions to fit standard fractional-inch dimension shafting. As they are usually contained in an assembled unit, they are specified by their nominal fractional-inch bore size.

Antifriction Bearings

All antifriction bearings consist of two hardened rings called the inner and outer rings, a separator, and hardened rolling elements which may be either balls or rollers. Bearing size is usually given in terms of what are called boundary dimensions. These are the *Outside Diameter*, the *Bore*, and the *Width*. The inner and outer rings provide continuous tracks or races for the balls or rollers to roll in. The separator or retainer properly spaces the rolling elements around the track and guides them through the load zone. Other words and terms used in describing antifriction bearings are the face, shoulders, corners, etc.

Terms used to describe roller bearings are a little different in that what is normally the outer ring is called the cup, and the inner ring the cone. The word cage is standard for taper roller bearings rather than separator or retainer.

The function of the separator, also called retainer or cage, is to properly space the rolling elements around the track and guide them through the load zone. As the bearing rotates and the rolling elements roll in the race, the separator rides with them. It is the weakest point in an antifriction bearing as sliding friction is always present between the separator pockets and the rolling elements.

Ball Bearing Dimensions

Basic type bearings for general use in industry are manufactured to standardized dimensions of bore, outside diameter, and width (boundary dimensions). Tolerances for these critical dimensions and for the limiting dimensions for corner radii have also been standardized. Therefore, all types and sizes of ball bearings made to standardized specifications are satisfactorily interchangeable with other makes of like size and type.

Most basic ball bearings are available in four different "series" known as *extra light, light, medium,* and *heavy.* The names applied to each series are descriptive of the relative proportions and load-carrying capacities of the bearings. This means that there are as many as four bearings (one in each series) with the same bore size but with different widths, outside diameters and load-carrying capacities.

It is also possible to select as many as four bearings with the same outside diameter (one in each series) with four different bore sizes, widths, and load-carrying capacities. Thus there is a choice of four different shaft sizes without changing the diameter of the housing.

Bearing manufacturers designate the various series by using numbers which they incorporate into their basic numbering systems. The extra light series is designated as the "100" series the light as "200," the medium as "300," and the heavy as the "400" series.

Ball Bearing Numbering Systems

Antifriction ball bearings were first manufactured on a large scale in Europe where the metric system of measurement is used. When the manufacture of ball bearings was started in America the practice of using metric sizes was continued. Because of this, metric ball bearing sizes are interchangeable throughout the western world. However, the fact they are made to even metric sizes results in dimensions that have no relation to inch fractions when converted to the inch system of measurement. For example, a ball bearing with a 60 millimeter bore measurement when converted to the inch system measures 2.3622 inches. Conversion of millimeter dimensions to inch dimensions is accomplished by multiplying the millimeter dimension by .03937, which is the equivalent of one millimeter in thousandths of an inch. For approximate conversion purposes one millimeter is roughly 1/25 of an inch.

The basic ball bearing number is made up of three digits. The first digit indicates the bearing series i.e. "100," "200," "300," or "400." The second and third digit, from 04 up, when multiplied by 5 indicates the bearing bore in millimeters. An example in each of the four duty series would be:

Basic Number	Duty Series	Bore In mm's	Bore In Inches
108	Extra Light	40	1.5748
205	Light	25	0.9843
316	Medium	80	3.1496
420	Heavy	100	3.9370

Ball bearings having a basic number under 04 have the bore dimensons listed below:

Basic Number	Bore In mm's	Bore In Inches
00	10	0.3937
01	12	0.4727
02	15	0.5906
03	17	0.6693

Cylindrical Roller Bearings (Numbers)

One of the most widely used types of roller bearings is one having rollers that are approximately equal in length and diameter. This specific type, which is called a "straight" roller bearing by some manufacturers, is made to the same standard dimensions as ball bearings. The same basic numbering system is used, although manufacture is limited to the light 200 series and the medium 300 series. This type roller bearing is interchangeable with ball bearings of like size and series.

Standard Ball Bearing Sizes

The basic three digit ball bearing number indicates the bearing duty series and the bearing bore in millimeters. All standard ball bearings in any of the four duty series having the same last two digits in their number, have the same diameter bore. Listed in Table 12 are the standard ball bearing bore sizes in nominal millimeters and equivalent decimal inches. Tables listing the boundary dimensions for standard size bearings in the "100," "200," "300," and "400" series are included in the appendix.

Table 12. Standard Ball Bearing Bore Sizes

Basic Bearing # Last 2 Digits	mm Bore Size	Inch Bore Size	Basic Bearing # Last 2 Digits	mm Bore Size	Inch Bore Size
00	10	0.3937	18	90	3.5433
01	12	0.4724	19	95	3.7402
02	15	0.5906	20	100	3.9370
03	17	0.6693	21	105	4.1339
04	20	0.7874	22	110	4.3307
05	25	0.9843	24	120	4.7244
06	30	1.1811	26	130	5.1181
07	35	1.3780	28	140	5.5118
08	40	1.5748	30	150	5.9055
09	45	1.7717	32	160	6.2992
10	50	1.9685	34	170	6.6929
11	55	2.1654	36	180	7.0866
12	60	2.3622	38	190	7.4803
13	65	2.5591	40	200	7.8740
14	70	2.7559	42	210	8.2677
15	75	2.9528	44	220	8.6614
16	80	3.1496	48	240	9.4488
17	85	3.3465			

Spherical Roller Bearings

The double-row spherical roller bearing is a self-aligning bearing utilizing rolling elements shaped like barrels. The outer ring has a single spherical raceway. The double-shoulder inner ring has two spherical races separated by a center flange. The rollers are retained and separated by an accurately constructed cage.

This type of bearing is inherently self-aligning because the assembly of the inner unit (ring, cage, and rollers) is free to swivel within the outer ring. Thus there is automatic adjustment which allows successful operation under severe misalignment conditions. It will support a heavy radial load and heavy thrust loads in both directions.

Generally speaking, bearing applications have a rotating inner ring and a stationary outer ring. When correctly assembled the inner ring is sufficiently tight on the shaft to ensure that both inner ring and shaft turn as a unit and "creeping" of the ring on the shaft does not occur. Should creeping occur, there will be overheating, excessive wear, and erosion between the shaft and the inner ring. For normal applications the inner ring is press-fitted to the shaft and/or clamped against a shoulder with a locknut. However, on applications subjected to severe shock or unbalanced loading the usual press fit or locknut clamping does not grip tightly enough to prevent creeping. For such applications a design providing maximum grip of the inner ring on the shaft is required. In providing this tremendous grip two very important conditions must be controlled. First the stress in the inner ring must remain below the elastic limit, and second the internal bearing clearance must not be eliminated. In addition, there must be a practical mounting method.

The taper bore self-aligning spherical roller bearing incorporates the features required. Size for size its capacity is greater than any other type of bearing. The taper bore provides a simple method of mounting which allows controlled stretching of the inner ring to obtain maximum gripping power. When mounting this style of bearing the inner ring is forced on the taper by tightening a locknut. As the inner race stretches, and its grip increases, the internal bearing clearance is reduced. The bearing is manufactured with sufficient internal clearance to allow this stretching of the inner ring. The grip and the clearance are controlled by checking internal clearance before mounting, and tightening the nut sufficiently to reduce the internal clearance by a specific amount.

Another feature that the taper bore bearing makes possible is the adapter-sleeve style of mounting. The use of a tapered adapter sleeve allows mounting the taper bore bearing on straight cylindrical surfaces.

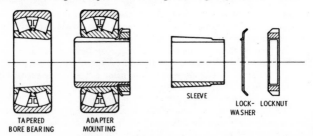

TAPERED
BORE BEARING   ADAPTER
MOUNTING   SLEEVE   LOCK- LOCKNUT
WASHER

Fig. 112

It also provides an easy means of locating the bearing. In many cases this feature is the reason for using the taper bore bearing for applications where loading is relatively light.

Installation Of Taper Bore Style—Before mounting spherical taper-bore roller bearings on taper-shaft fits or adapters, the internal clearance should be checked and recorded. The measurement is made on one side of the bearing only. Recommended practice is to rest the bearing upright on a table and insert the feeler gage between the top roller and the inside of the outer ring.

Mounting Procedure

1. Check internal clearance between the rollers and the outer race on the open side. Feeler gauge must be inserted far enough to contact entire roller surface.
2. Lightly oil surface of bearing bore and install on taper shaft or taper adapter sleeve *without* lock washer.
3. Check internal clearance as nut is tightened. Nut must be tightened until internal clearance is reduced the amount shown in the table below.
4. Remove nut, install lock washer and retighten nut. Secure lock washer.

Shaft Dia.	Clearance Reduction
1⅝ to 3½	.001 to .002
3½ to 6½	.002 to .0035
6½ to 10¼	.0035 to .0055

The above values are for most common applications. For special cases such as high temperature equipment, bearings on hollow shafts with steam passing through etc., consult manufacturers specifications and instructions for special applications.

Taper Roller Bearings

Taper roller bearings are a separate group of roller bearings because of their design and construction. They consist fundamentally of tapered cone-shaped rollers operating between tapered raceways. They are so constructed, and the angle of all rolling elements so proportioned, that if straight lines were drawn from the tapered surfaces of each roller and raceway they would meet at a common point on the center line of the axis of the bearing as illustrated in Fig. 113.

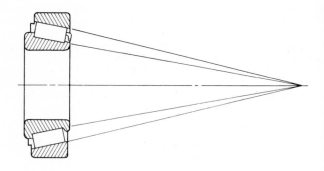

Fig. 113

One major difference between taper roller bearings (separate types), and most antifriction bearings, is that they are adjustable. In many cases this is a distinct advantage as it permits accurate control of bearing running clearance. The proper amount of running clearance may be assembled into the bearing at installation to suit the specific application. This feature also permits preloading the bearing for applications where extreme rigidity is required. This adjustable feature also requires that proper procedure be followed at assembly to ensure correct setting. Generally the best setting is one of minimum clearance allowing free running with no appreciable end play.

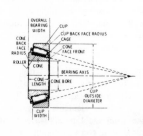

Fig. 114

The basic parts of a taper roller bearing are the *cone* or inner race, the *taper rollers,* and the *cage* (which is called the retainer or separator), and the *cup* or outer race.

The most widely used taper roller bearing is a single-row type with cone, rollers, and cage factory-assembled into one nuit, with the cup independent and separable. Many additional styles of single and multiple row taper roller bearings are made, including unit assemblies factory preadjusted.

Taper Roller Bearing Adjustment—While each single row taper bearing is an individual unit, their construction is such that they must be mounted in pairs so that thrust may be carried in either direction. Two systems of mounting are employed "direct" and "indirect" as shown in Fig. 115.

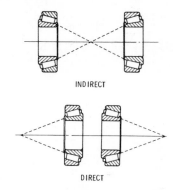

INDIRECT

DIRECT

Fig. 115

Indirect mounting is used for applications where maximum stability must be provided in a minimum width.

Direct mounting is used where this assembly system offers mounting advantages and maximum stability is not required or desired.

The terms *face to face* and *back to back* may also be used to describe these systems of mounting. If these terms are used they must be qualified by stating whether the terms are used in regard to the cup or the cone. Indirect and direct mounting may more appropriately be referred to as *cone-clamped* and *cup-clamped*. Cone-clamped describes indirect mounting which is normally secured by clamping against the cone. Cup-clamped describes direct mounting which is normally secured by clamping against the cup.

Many devices are used for adjusting and clamping taper roller bearing assemblies. The following six illustrations are the basic ones, others are in general, variations of these.

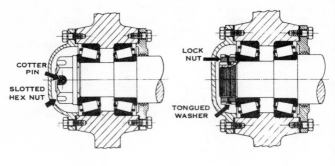

Fig. 116   Fig. 117

1. The slotted hex nut and the cotter pin are used to adjust bearings in implement wheels and automotive front wheels. Simply tighten the nut while rotating the wheel, until a slight bind is obtained in the bearing. That ensures proper seating of all parts. Then, back off the nut one slot, lock with a cotter pin. Result is free-running clearance in the bearing. Bearing is rotated in order to seat rollers against the cone rib—should be done when making any bearing adjustment.

2. Here two standard lock nuts and a tongued washer are used to adjust the bearing. They provide a much finer adjustment than a slotted hex nut. Pull up the inner nut until there is a slight bind on the bearings then back off just enough to allow running clearance after the outer or "jamb" nut is tightened. This type of adjustment device is used in full floating rear axles and industrial applications where slotted hex nuts are not practicable or desirable.

Fixed and Floating Bearings

Temperature variations will expand and contract the components of any machine. Because of this, it is essential that such parts be permitted to expand and contract without restriction. For that reason, only one bearing on any one shaft should be fixed axially in the housing (called a *fixed* or *held* bearing) to prevent axial or endwise motion. All other bearings on that same shaft should have adequate axial clearance in the housing (referred to as *floating* or *free* bearings).

Basic Adjusting Devices

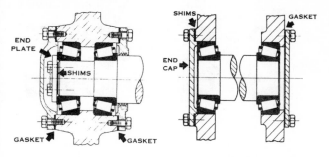

Fig. 118

Fig. 119

3. In this case, shims are used between the end of the shaft and the end plate. The shim pack is selected to give the proper setting of the bearing. This will vary with the unit in which the bearings are used. The end plate is held in place by cap screws. The cap screws are wired together for locking. A slot may be provided in the end plate to measure the shim gap.

4. Here is another adjustment using shims. They are located between the end cap flange and the housing. Select the shim pack which will give the proper bearing running clearance recommended for the particular application. End cap is held in place by cap screws which can be locked with lock washer (as shown) or wired. This easy method of adjustment is used where press-fitted cones are used with loose-fitted cups.

The illustration in Fig. 122 shows a typical fixed and floating bearing assembly. The fixed bearing is clamped securely in its housing. The floating bearing has clearance on both sides within the housing. Thus movement is allowed as changes in temperature cause the shaft to increase or decrease in length.

Generally it is preferred to hold the bearing at the drive end. However, consideration is sometimes given to bearing load, and the bearing carrying the smaller radial load is fixed. This is usually in cases where the fixed bearing is subjected to a thrust load of some magnitude and distributing the load in this manner tends to equalize it.

Another very important consideration in the determination of which bearing should be fixed, are operating clearances. When the clearances of components rotating with the shaft must be held to close tolerances, the closest bearing to the close clearance point should be fixed.

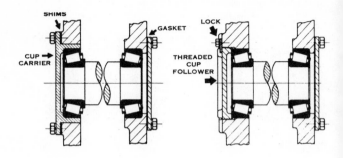

Fig. 120   Fig. 121

5. Shims are also used in this adjustment, as in No. 4. The difference here is that one cup is mounted in a carrier. This adjusting device is commonly used in gear boxes and drives. It is also used in industrial applications for ease of assembly or disassembly.

6. The threaded cup follower shown above is another common adjusting device. It is used in gear boxes, drives or automotive differentials. The follower is locked by means of a plate and cap screws. The plate fits between the lugs of the follower.

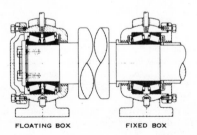

FLOATING BOX   FIXED BOX

Fig. 122

In gear reducers where herringbone gears are used, only one bearing on one of the shafts should be fixed, usually the output shaft. In this style of assembly the "V" shape of the gear teeth will locate the mating gear and shaft axially.

When a flexible coupling is used to connect two shafts, a fixed bearing is required on each end of the shafts, as a flexible coupling permits endwise motion of both shafts.

Antifriction Bearing Care & Handling—The principal reason for antifriction bearing failure is the entrance of dirt or grit. Second is mechanical damage from improper handling. The two most important rules therefore, in handling antifriction bearings are:

1. Keep bearings and parts clean.
2. Apply force to the tight fitting ring only.

PUMPS

Pumps are broadly classified with respect to their construction, or the service for which they are designed. The three groups into which most pumps in common use fall are: *centrifugal, reciprocating,* and *rotary.*

Centrifugal Pumps

Centrifugal force, from which this pump takes its name, acts upon a body moving in a circular path tending to force it farther from the center of the circle.

Inside the body of a centrifugal pump the impeller forces the liquid to revolve and generate centrifugal force. The impeller blades, as shown at right, are usually curved backwards with reference to the direction of rotation. The liquid is drawn in through the center or "eye" of the impeller, and it is whirled around by the blades, it is thrown outward by the centrifugal force and passes through the discharge outlet.

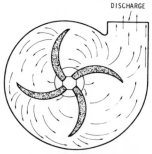

Fig. 123

There are two principal classes of centrifugal pumps, single-stage and multistage. The single-stage pump has a single impeller. By arranging a number of centrifugal pumps in series so the discharge of one is led to the suction of the succeeding pump, the head or pressure may be multiplied as required. Multistage pumps are made with a common housing, and internal passages so arranged that liquid flows from the discharge of one stage to the inlet of the next.

Reciprocating Pumps

The reciprocating pump has a back-and-forth motion as the pumping element alternately moves forward and backward. It moves liquid by displacing the liquid with a solid, usually a piston or plunger. The principle of operation is called *positive displacement.*

The piston pumping element is a relatively short cylindrical part that is moved back and forth in the pump chamber, or cylinder. The distance that the piston travels back and forth, called the *stroke,* is generally greater than the length of the piston. Leakage past the piston is usually controlled by packings or piston rings. The piston in normal operation moves back and forth within the cylinder.

A *plunger* pumping element is generally longer than the stroke of the pump. In operation the plunger moves into and withdraws from the cylinder. To prevent leakage past the plunger, packings are contained in the end of the cylinder through which the plunger moves.

As the pumping element in a reciprocating pump travels to and fro, liquid is alternately moved into the pump chamber and moved out. The period during which the element is withdrawing from the chamber and liquid is entering is called the *suction,* or *intake* stroke. Travel in the opposite direction during which the element displaces the liquid is called the *discharge* stroke. Check valves are placed in the suction and discharge passages to prevent backflow of the liquid. The valve in the suction passage is opened and the discharge passage valve is closed during the suction stroke. Reversal of liquid flow on the discharge stroke causes the suction valve to close and the discharge valve to open.

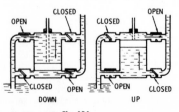

Fig. 124

Fig. 124 illustrates the position of the valves during travel in each direction. At one end of each cylinder the suction valves are open to admit liquid and the discharge valves closed to prevent backflow from the discharge passage. On reversal of direction, the suction valves are closed to prevent backflow into the suction passage and liquid moves out through the open discharge valves.

Rotary Pumps

Rotary pumps are also positive-displacement type pumps in operation. As their flow is continuous in one direction, no check valves are required. Different designs make use of such elements as vanes, gears, lobes, cams, etc., to move the material The principle of operation is similar with all of these elements in that the element rotates within a close fitting casing which contains the suction and discharge connections (Fig. 125).

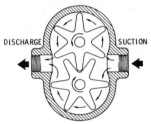

Fig. 125

At the pump suction port the liquid enters chambers formed by spaces in the elements, or between the surface of the elements and the internal chamber surface. The liquid is carried with the elements as they rotate, and it is literally squeezed out the discharge as the elements mesh or the volume of the chambers are reduced to practically zero.

Rotary pumps have close running clearances and generally are self-priming. In operation they produce a very even continuous flow with almost no pulsation. The delivery capacity is constant regardless of pressure, within the limits of operating clearances and power.

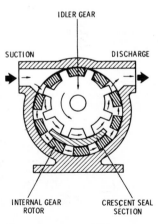

Fig. 126

Table 13. Troubleshooting

Trouble Sympton	Key to Causes
Pump fails to discharge	1, 2, 3, 4, 6, 7, 12
Pump discharges then stops	4, 8, 9, 10, 11
Pump not up to capacity	1, 2, 4, 7, 8, 12, 13, 15
Pump noisy or vibrates	7, 11, 12, 14, 16, 17, 18, 19, 23
Pump takes too much power	5, 14, 17, 20, 21, 22, 24

Trouble Causes

1. Suction or discharge valves closed
2. Direction of rotation wrong
3. Lift too high
4. Supply level low
5. Supply level high
6. Pump not primed
7. Blocked lines
8. Leaks in stuffing box
9. Leaks in suction line
10. Vents blocked
11. Air or vapor in liquid
12. Impeller or rotor damaged
13. Speed too low
14. Misalignment
15. Pump worn or excessive clearance
16. Shaft bent
17. Binding of rotating elements
18. Cavitation
19. Worn or defective pump or motor bearings
20. Motor undersize
21. Specific gravity of material changed
22. Viscosity of material changed
23. Relief valve chattering
24. Stuffing box tight

STRUCTURAL STEEL

American Standard Angles

The symbol used to indicate an angle shape is (∠). The usual method of billing is to state the symbol, then the long leg, the short leg, the thickness and finally the length. For example, ∠ 6 × 4 × ⅜ × 12'4". When the legs are equal, both lengths are stated.

Fig. 127

American Standard Beams

Called "I" beams because their resemblance to the capital letter "I." The symbol used to indicate the beam shape is the letter (I). The usual method of billing is to state the depth, the symbol, the weight per foot, and finally, the length. For example, 15 I 42.9 × 18'4½".

American Standard Channels

May be compared to an I beam that has been trimmed on one side to give a flat back web. The symbol used to indicate the standard channel is ([). The usual method of billing is to state the depth, the symbol, the weight per foot and finally the length. For example, 10 [15.3 × 16'6".

American Standard Wide Flange Beams or Columns

Also referred to as "B," "CB," or "H" shapes. May also be compared to an I beam with extra wide flanges. The symbol used to indicate the standard wide flange shape is (WF). The usual method of billing is to state the depth, the symbol, the weight per foot and finally the length. For example, 12 45 × 24'8".

Fig. 128

Fig. 129

Fig. 130

Structural steel is produced at rolling mills in a wide variety of standard shapes and sizes. In this form it is referred to as "plain material." In addition to the plates and bar stock, the four shapes illustrated above are the most widely used. Other standard shapes produced are *Tee's, Zee's, Rails,* and various special shapes. All standard structural shapes are made to a standardized series of nominal sizes. Within each size group there is a wide range of weights and dimensions.

Simple Square-Framed Beams

Square-framed beams are, as the name implies, beams that intersect or connect at right angles. This is the commonest type of steel construction. Two types of connections may be used in framed construction—*framed* and *seated*. In the framed type, shown in Fig. 131A the beam is connected by means of fittings (generally a pair of angle irons) attached to its web. With the seated connection (Fig. 131B) the end of the beam rests on a ledge or seat.

Clearance Cuts

When connecting one member to another it is often necessary to notch or cut away both flanges of the entering member to avoid flange interference. Such a notch is called a *cope*, a *block*, or a *cut*. The term "cope" is usually used if the cut is to follow closely the shape of the member into which it will fit. When the cut is rectangular in shape with generous clearance, it is usually called a "block-out." Unless there is some reason for a close matching fit the "block-out" is recommended as it is the easiest and most economical notch to make.

When making block-out cuts the dimensions of the rectangular notch may be obtained from tables of structural steel dimensions. The tables list the "K" and "a" dimensions for various size and weight members. The "K" and "a" values determine the dimensions of the cut as they indicate the maximum points of interference. While the notch is made in the entering member, the values from the table for the supporting member determine the notch dimensions. The steel must be cut to length before the block-out cut is made.

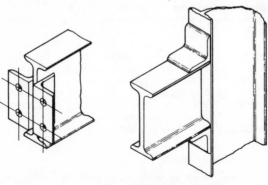

A   **Fig. 131**   B

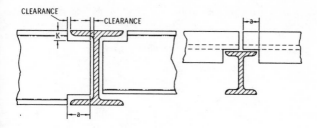

Fig. 132

Square-Framed Connections (Two-Angle Type)

"Standard" connections should be used when fabricating structural steel members to ensure proper assembly with supporting members at installation. Some of the terms, dimensions etc., of standardized connections are the following:

Spread—The distance between hole centers in the web of the supporting member, and the holes in the *attached* connection angles is called the *spread* of the holes. The spread dimension is standardized at 5½ inches as shown below.

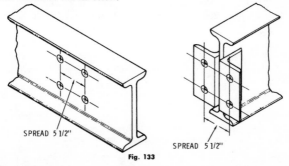

Fig. 133

Angle Connection Legs

The legs of the angles used as connections are specified according to the surface to which they are connected, as shown below. The legs which attach to the entering steel to make the connection, are termed *web* legs. The legs of the angles that attach to the supporting member are termed *outstanding* legs. The lines on which the holes are placed are called *gauge* lines. The distance between gauge lines, or from a gauge line to a known edge, are called *gauges*.

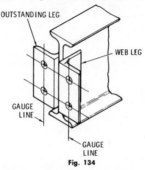

Fig. 134

Fabrication Terms

Commonly used structural steel fabrication terms are illustrated in Fig. 135. Use of these terms, because they are descriptive, aids in understanding of steel fabrication and reduces the probability of errors.

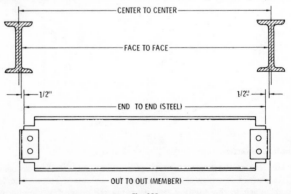

Fig. 135

Steel—Various structural steel shapes and forms.

Member—An assembly of a length of steel and its connection fittings.

Center-to-Center—The distance from the center line of one member to the center line of another member.

Face-to-Face—The distance between the facing web surfaces of two members. It is the center-to-center distance minus the ½-web thickness of each member.

End-to-End—The overall length of the steel. Should be 1 inch shorter than the opening into which it will be placed. The 1-inch clearance is provided for assembly and as an allowance for inaccuracies.

Out-to-Out—The overall length of the assembled member. To provide assembly clearance the connection angles are positioned on the member so the out-to-out distance before assembly is slightly less than the face-to-face distance.

Connection Hole Locations

The various terms and the constant dimension for standard square-framed two-angle type connections are illustrated in Fig. 136.

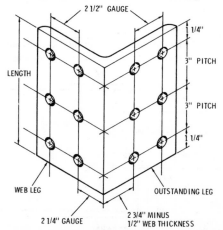

Fig. 136

Web-Leg Gauge—The distance from the heel of the angle to the first gauge line on the web leg is called "web-leg gauge." This dimension is constant as it is standardized at 2¼ inches.

Outstanding-Leg Gauge—The distance from the heel of the angle to the first gauge line on the outstanding leg is called the "outstanding-leg gauge." This dimension varies as the thickness of the web of the member varies in order to maintain a constant 5½-inch spread dimension. The outstanding leg-gauge dimension is determined by subtracting the web thickness from 5½ inches and dividing by two. A simpler way to determine the outstanding-leg dimension is to subtract the ½-web thickness from 2¾ inches which is ½ the spread.

Gauges—The distance between gauge lines, or from a gauge line to a known edge, are called "gauges." When more than one row of holes are used, the gauge is 2½ inches. This dimension is constant.

Pitch—The distance between holes on any gauge line is called "pitch." This dimension is standardized at 3 inches.

End Distance—The end distance is equal to one half of the remainder left after subtracting the sum of all pitches from the length of the angle. By common practice the angle length is selected to give a 1¼-inch end distance.

Dimensions of Two-Angle Connections

Detail drawings for fabrication of structural steel members specify dimensions, hole locations, etc. A typical detail drawing for a two-

2 ∠   4"x 3 1/2"x 3/8" x 8 1/2" 2 1/4"

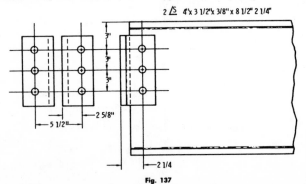

Fig. 137

angle connection is illustrated below. Note that the angles are slightly above center, the uppermost hole being 3 inches below the top of the beam. Whenever practicable, the uppermost holes are set at 3 inches below the top of the beam as it makes for standardization and tends to reduce errors in matching connections.

The selection of steel sizes is usually done by persons competent to make strength calculations based on anticipated loadings. Connection details however, may not be provided and the selection of connections may be left to the fabricator. To aid in the selection of connections the following Table 14 of two-angle connections and two-angle connection dimensions are provided.

For purposes of simplification the connections have been given numbers 1 through 6. While these are not standard connection numbers, dimensions conform to connection standards. Note also that only connections using ¾-inch diameter bolts are shown. This is also for the purposes of simplification as ¾-inch bolts are adequate for use with the steel sizes for which these connections are used.

To select angle connections, find the size in the first column of the appropriate table, and the weight per foot in the second column. Follow along the weight line to the column that corresponds to the span length. The number in this column indicates the connection angle to use. The dimensions to which the selected angles should be fabricated are given under Two-Angle Connection Dimensions.

Steel Elevation

Unless otherwise stated on the drawing, all square framed beam members are presumed to be parallel or at right angles to one another. Also that their webs are in a vertical plane and that they are in a level position end to end. Elevation information is usually given by note, stating the vertical distance above some established horizontal plane.

Table 14. Two-Angle Connections
For Uniformly Loaded Beams and Channels

Size Inches	Weight Lbs/Ft	Span In Feet												
		4	6	8	10	12	14	16	18	20	22	24	26	28
24	120									6	6	6	6	6
	100							6	6	6	6	6	6	6
	79.9							6	6	6	6	6	6	6

Table 14. Two-Angle Connections
For Uniformly Loaded Beams and Channels (Continued)

Size Inches	Weight Lbs/Ft	Span In Feet												
		4	6	8	10	12	14	16	18	20	22	24	26	28
20	95							5	5	5	5	5	5	5
	65.4					5	5	5	5	5	5	5	5	5
18	70					4	4	4	4	4	4	4	4	4
	54.7					4	4	4	4	4	4	4	4	4
15	50			4	4	4	4	4	4	4	4	4	4	4
	42.9			4	4	4	4	4	4	4	4	4	4	4
12	50			3	3	3	3	3	3	3	3	3	3	3
	31.8			3	3	3	3	3	3	3	3	3	3	3
10	35			2	2	2	2	2	2	2	2	2	2	2
	25.4		2	2	2	2	2	2	2	2	2	2	2	2
8	23	2	2	2	2	2	2	2	2	2	2	2	2	2
	18.4	2	2	2	2	2	2	2	2	2	2	2	2	2
7	20	1	1	1	1	1	1	1	1	1	1	1	1	1
	15.3	1	1	1	1	1	1	1	1	1	1	1	1	1
6	17.2	1	1	1	1	1	1	1	1	1	1	1	1	1
	12.5	1	1	1	1	1	1	1	1	1	1	1	1	1
5	14.7	1	1	1	1	1	1	1	1	1	1	1	1	1
	10	1	1	1	1	1	1	1	1	1	1	1	1	1

For Uniformly Loaded Wide Flange Beams

Size Inches	Lbs/Ft	Span In Feet												
		4	6	8	10	12	14	16	18	20	22	24	26	28
24	120												6	6
	76						6	6	6	6	6	6	6	6
21	96						6	6	5	5	5	5	5	5
	62					5	5	5	5	5	5	5	5	5
18	70					5	5	5	5	5	4	4	4	4
	50					4	4	4	4	4	4	4	4	4
16	96									4	4	4	4	4
	64						4	4	4	4	4	4	4	4
	36				4	4	4	4	4	4	4	4	4	4
14	38						3	3	3	3	3	3	3	3
	34					3	3	3	3	3	3	3	3	3
	30					3	3	3	3	3	3	3	3	3
12	36				3	3	3	3	3	3	3	3	3	3
	31					3	3	3	3	3	3	3	3	3
	27			3	3	3	3	3	3	3	3	3	3	3

Table 14. Two-Angle Connections
For Uniformly Loaded Beams and Channels (Continued)

Size Inches	Weight Lbs/Ft	Span In Feet											
		4	6	8	10	12	14	16	18	20	22	24	26	28
10	29			2	2	2	2	2	2	2	2	2	2	2
	25			2	2	2	2	2	2	2	2	2	2	2
	21		2	2	2	2	2	2	2	2	2	2	2	2
8	20	2	2	2	2	2	2	2	2	2	2	2	2	2
	17	2	2	2	2	2	2	2	2	2	2	2	2	2

Note

Connections must not be used for shorter spans than indicated as their capacity does not equal that of the steel in shorter spans.

TWO-ANGLE CONNECTION DIMENSIONS

$\frac{13}{16}$" Holes For $\frac{3}{4}$" Bolts

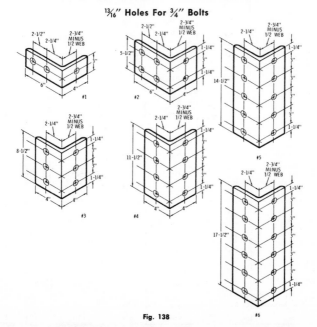

Fig. 138

Table 15. Standard Channels

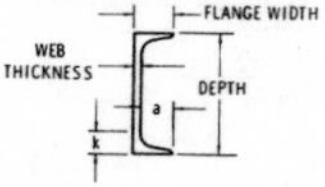

Depth	Weight Per Ft.	Flange Width	Web Thick.	a	k
3″	4.1	1⅜	3/16	1¼	⅝
	5.0	1½	¼	1¼	⅝
	6.0	1⅝	⅜	1¼	⅝
4″	5.4	1⅝	3/16	1⅜	⅝
	7.25	1¾	5/16	1⅜	⅝
5″	6.7	1¾	3/16	1½	11/16
	9.0	1⅞	5/16	1½	11/16
6″	8.2	1⅞	3/16	1¾	¾
	10.5	2	5/16	1¾	¾
	13.0	2⅛	7/16	1¾	¾
7″	9.8	2⅛	¼	1⅞	13/16
	12.25	2¼	5/16	1⅞	13/16
	14.75	2¼	7/16	1⅞	13/16
8″	11.5	2¼	¼	2	13/16
	13.75	2⅜	5/16	2	13/16
	18.75	2½	½	2	13/16
9″	13.4	2⅜	¼	2¼	⅞
	15.0	2½	5/16	2¼	⅞
	20.0	2⅝	7/16	2¼	⅞
10″	15.3	2⅝	¼	2⅜	15/16
	20.0	2¾	⅜	2⅜	15/16
	25.0	2⅞	9/16	2⅜	15/16
	30.0	3	11/16	2⅜	15/16
12″	20.7	3	5/16	2⅝	1 1/16
	25.0	3	⅜	2⅝	1 1/16
	30.0	3⅛	½	2⅝	1 1/16
15″	33.9	3⅜	7/16	3	1 5/16
	40.0	3½	9/16	3	1 5/16
	50.0	3¾	¾	3	1 5/16
18″	42.7	4	7/16	3½	1 5/16
	45.8	4	½	3½	1 5/16
	51.9	4⅛	⅝	3½	1 5/16
	58.0	4¼	11/16	3½	1 5/16

Table 16. Standard Beams

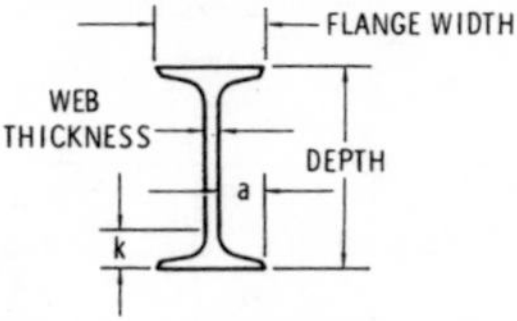

Depth	Weight Per Ft.	Flange Width	Web Thick.	a	k
3″	5.7	2⅜	3/16	1⅛	9/16
	7.5	2½	⅜	1⅛	9/16
4″	7.7	2⅝	3/16	1¼	⅝
	9.5	2¾	5/16	1¼	⅝
5″	10.0	3	¼	1⅜	11/16
	14.75	3¼	½	1⅜	11/16
6″	12.5	3⅜	¼	1½	¾
	17.75	3⅝	½	1½	¾
7″	15.3	3⅝	¼	1¾	13/16
	20.0	3⅞	7/16	1¾	13/16
8″	18.4	4	5/16	1⅞	⅞
	23.0	4⅛	7/16	1⅞	⅞
10″	25.4	4⅝	5/16	2⅛	1
	35.0	5	⅝	2⅛	1
12″	31.8	5	⅜	2⅜	1⅛
	35.0	5⅛	7/16	2⅜	1⅛
12″	40.8	5¼	½	2⅜	1 5/16
	50.0	5½	11/16	2⅜	1 5/16
15″	42.9	5½	7/16	2½	1¼
	50.0	5⅝	9/16	2½	1¼
18″	54.7	6	½	2¾	1⅜
	70.0	6¼	½	2⅞	1 9/16
20″	65.4	6¼	⅝	2⅞	1 9/16
	75.0	6⅜	¾	2¾	1⅜
20″	85.0	7	11/16	3¼	1¾
	95.0	7¼	13/16	3¼	1¾

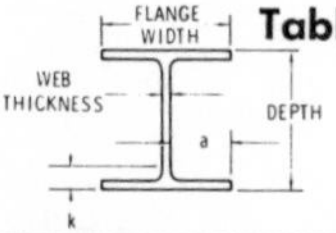

Table 17. Wide Flange—CB Sections

Nom. Depth	Weight Per Ft.	Flange Width	Web Thick.	a	k
4″	10	4	1/4	1 7/8	7/16
5″	16	5	1/4	2 3/8	5/8
6″	12	4	1/4	1 7/8	9/16
	15.5	6	1/4	2 3/8	9/16
8″	13	4	1/4	1 7/8	9/16
	17	5 1/4	1/4	2 1/2	5/8
	20	5 1/4	1/4	2 1/2	11/16
	24	6 1/2	1/4	3 1/8	13/16
	28	6 1/2	5/16	3 1/8	13/16
	31	8	5/16	3 7/8	13/16
	35	8	5/16	3 7/8	7/8
	40	8 1/8	3/8	3 7/8	15/16
	48	8 1/8	7/16	3 7/8	1 1/16
	58	8 1/4	1/2	3 7/8	1 3/16
	67	8 1/4	9/16	3 7/8	1 5/16
10″	15	4	1/4	1 7/8	9/16
	21	5 3/4	1/4	2 3/4	11/16
	25	5 3/4	1/4	2 3/4	13/16
	29	5 3/4	5/16	2 3/4	7/8
	33	8	5/16	3 7/8	15/16
	39	8	5/16	3 7/8	1 1/16
	45	8	3/8	3 7/8	1 1/8
	49	10	3/8	4 7/8	1 1/16
	60	10 1/8	7/16	4 7/8	1 3/16
	72	10 1/8	1/2	4 7/8	1 5/16
	100	10 3/8	11/16	4 7/8	1 5/8
12″	27	6 1/2	1/4	3 1/8	13/16
	36	6 5/8	5/16	3 1/8	15/16
	40	8	5/16	3 7/8	1 1/8
	50	8 1/8	3/8	3 7/8	1 1/4
	58	10	3/8	4 7/8	1 1/4
	65	12	3/8	5 3/4	1 3/16
	106	12 1/4	5/8	5 3/4	1 9/16
14″	30	6 3/4	5/16	3 1/4	7/8
	48	8	3/8	3 7/8	1 3/16
	68	10	7/16	4 3/4	1 5/16
	84	12	7/16	5 3/4	1 3/8

TWIST DRILLS

Drill Terms

For general purpose drilling of steel, a twist drill point angle of 118 degrees is generally recommended. For hard and tough materials, and for field work using a drill motor, a point angle of 135 degrees is recommended.

The heel or clearance angle for the 118-degree point should be about 8 to 12 degrees. The 135-degree point should have a clearance angle of about 6 to 9 degrees.

A twist drill cuts by wedging under the material and raising a chip. The steeper the point and the greater the clearance angle, the easier it is for the drill to penetrate. The blunter the point and the smaller the lip clearance, the greater is the support for the cutting edges. Thus, the greater point angle and lesser lip clearance for hard and tough materials, and the decreased point angle and increased lip clearance for softer materials.

When drilling some of the nonferrous metals there is a tendency for the drill point to bite in, or penetrate too rapidly. To overcome this, the cutting edge of the drill is slightly flattened in front as shown in the illustration for brass and copper in Fig. 140.

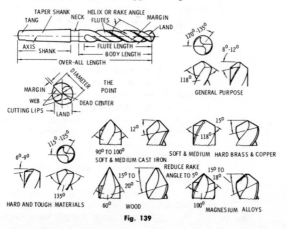

Fig. 139

ORIGINAL
MARGIN

CHISEL
EDGE

POINT OF DRILL AFTER
DRILL HAS BEEN CUT
BACK IN USE AND
REPOINTED

HOLD ORIGINAL
THICKNESS

A

USUAL METHOD OF THINNING
THE POINT OF A DRILL WHEN
THE WEB HAS BECOME TOO
THICK BECAUSE OF REPEATED
RE-POINTING

CUTTING LIP

CHISEL
EDGE

SURFACE
OF POINT   CUTTING LIP

ANGLES AND LENGTHS OF
CUTTING LIPS MUST BE
EQUAL

Fig. 140

Drill Sharpening

A new twist drill starts to wear as soon as it is placed in operation. Wear starts as a dulling along the cutting edges or lips and also a slight rounding of the corners as shown at the right. These dulled edges result in heat generation and a faster rate of wear which tends to extend back along the margins. When regrinding a twist drill all of this worn section must be removed. Sharpening the edges or lips only, without removing the worn margins, will not properly recondition a twist drill.

Most drills are made with webs which increase in thickness toward the shank. Therefore, after several sharpenings and shortenings of a twist drill the web thickness at the point increases, resulting in a longer chisel edge as shown at right. When this occurs it is necessary to reduce the web so that the chisel edge is restored to its normal length. This operation is called *web-thinning*.

Several different types of web-thinning are in common use. The method shown at right is perhaps the most common. The length A is usually made about ½ to ¾ the length of the cutting lip.

After the worn portion of the drill has been removed and the web thinned if necessary, the surfaces of the point must be reground. These two conical surfaces intersect with the faces of the flutes to form the cutting lips, and with each other to form the chisel edge. As in the case of any other cutting tool, the surface back of these cutting lips must not rub on the work, but must be relieved in order to permit the cutting edge to penetrate. Without such relief the drill could not penetrate the metal, but would only rub around and around.

In addition to grinding the conical surfaces to give the correct point angle and cutting clearance, both surfaces must be ground alike. Regardless of the point angle, the angles of the two cutting lips (A1 and A2) must be equal. Similarly the lengths of the two lips (L1 and L2) must be equal. Drill points of unequal angles or lips of unequal lengths will result in one cutting edge doing most of the cutting. This type of point will cause oversize holes, excessive wear and short drill life.

Drill Sharpening

To maintain the necessary accuracy of drill point angles, lip-lengths, lip-clearance angle, and chisel edge angle, the use of machine point-grinding is recommended. However, the lack of a drill-point grinding machine is not sufficient reason to excuse poor drill points. Drills may be pointed accurately by hand if proper procedure is followed and care exercised.

FREEHAND DRILL POINT GRINDING

1. Adjust grinder tool rest to a convenient height for resting back of forehand on it while grinding drill point.
2. Hold drill between thumb and index finger of left hand. Grasp body of drill near shank with right hand.
3. Place forehand on tool rest with centerline of drill making desired angle with cutting face of grinding wheel (Fig. 141A) and slightly lower end of drill (Fig. 141B).
4. Place heel of drill lightly against grinding wheel. Gradually raise shank end of drill while twisting drill in fingers in a counterclockwise rotation and grinding conical surfaces in the direction of the cutting edges. Exert only enough pressure to grind the drill point without overheating. Frequently cool drill in water while grinding.
5. Check results of grinding with a gauge to determine if cutting edges are the same lengths and at desired angle, and that adequate lip clearance has been provided.

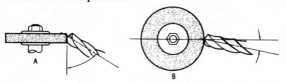

Fig. 141

STAIR LAYOUT

Knowledge and understanding of certain terms and practices helps considerably in the layout and installation of simple stairs. This is true for stairs as simple as a few steps to a full floor level of stairs, both straight and platform type.

Straight Stairs   *Platform Stairs*

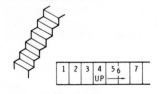

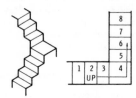

Fig. 142

Terms

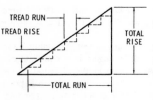

Total Rise—The vertical distance from floor surface to floor surface.

Total Run—The straight length from first rise to final rise.

Tread Rise—Vertical height of one rise.

Tread Run—Distance from one rise to next rise.

Conditions

A stair layout usually must comply with certain fixed conditions and/or specific dimensions such as; height, size of opening, available space, direction of run, etc. In addition there are certain general conditions which the layout must meet, such as tread-rise dimension, tread-run dimension, etc.

While there is no standard tread-rise dimension as such, experience has shown that it should be in the area of 7 inches. Too large a rise results in a steep stairway, sometimes referred to as a "hard" stairway, because it is difficult to both ascend and descend. Also the tread rise must be proportional to the tread run, that is the tread run narrows as the tread rise increases and vice versa. There are numerous methods or rules to determine the dimensions of the tread rise and run. A very simple one, easy to remember and quite satisfactory for general use, is that the sum of the tread-rise and -run dimensions should total 17 inches. The tread-rise dimension being held as close as possible to the optimum 7 inches. Another rule in common use is that the run dimension plus twice the rise dimension should equal 24.

Dimensions

Specific layout dimensions for stairs are determined from the given or known dimensions and/or conditions, such as size of opening, height from floor to floor, available space, direction of run etc. The manner or method of layout therefore is dictated by the given or known information. For instance, the most frequently encountered situation is one where the total rise (distance from level to level) is known and a set of straight stairs are to be installed. If space is not limited, all necessary dimensions for layout of a comfortable set of stairs may be calculated from this one dimension.

EXAMPLE

A straight set of stairs are to be constructed and installed from one level to the level above. The vertical distance from floor level to floor level is 12 feet 8 inches.

Total Rise—Vertical distance between levels 12′ 8″, or 104″

Tread Rise—The calculation of the tread rise is done in three steps. First determine the approximate number of rises by dividing the total rise by 7. Second select a whole number that is close to the number calculated. This will be the number of rises or steps in the stairs. Third calculate the tread rise dimension by dividing the total rise measurement by the number of rises or steps selected.

Using the example stated above, 7 will divide into 104 almost 15 times, therefore the selection is between 14 and 15 rises. If 15 rises are selected the tread rise will be $6^{15}\!/_{16}″$. If 14 rises are selected the tread rise will be $7^{7}\!/_{16}″$.

Tread Run—The tread run is calculated by subtracting the tread-rise dimension from 17. As this is an approximate figure, the actual tread-run dimension is rounded off to the closest convenient fraction, for instance, for a $6^{15}/_{16}''$ tread rise a 10″ run may be used, while for a $7^{7}/_{16}''$ rise a 9½″ run would be proper.

Total Run—The total run is calculated by multiplying the tread-run dimension by *one less* than the number of rises. Using the figures from the above example, the total run for 14 rises would be 123½″ (13 × 9½), and the total run for 15 rises would be 140″ (14 × 10).

The two examples stated above are illustrated in Fig. 143.

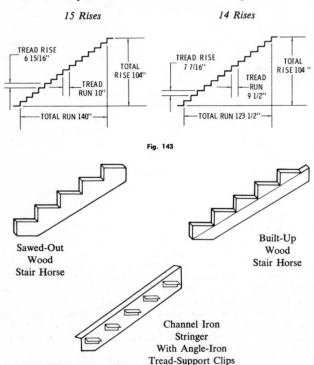

15 Rises   *14 Rises*

TREAD RISE
6 15/16″   TOTAL
                            RISE 104″
                 TREAD
                 RUN 10″

— TOTAL RUN 140″ —

TREAD RISE
7 7/16″   TOTAL
                               RISE 104″
                     TREAD
                     RUN
                     9 1/2″

— TOTAL RUN 123 1/2″ —

Fig. 143

Sawed-Out
Wood
Stair Horse

Built-Up
Wood
Stair Horse

Channel Iron
Stringer
With Angle-Iron
Tread-Support Clips

Fig. 144

Stairs are usually supported by either *stair horses* or by *stair stringers*. Stair horses are an underneath style of support member, with the stair tread resting on the steps of the horses. Stair stringers are side support members, the treads are located between the stringers and are supported by end attachment.

When constructing simple stairs, the stair horse style of construction us usually used with wood materials and the stair stringer style when structural steel materials are used.

Stair Length

The length of stair horses or stringers may be greater or less than the stair length, depending on the construction that is used. The actual length of the horse or stringer is determined during the layout. A few construction styles are shown in Fig. 145.

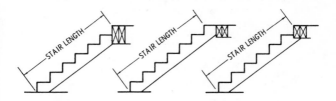

Fig. 145

The usual practice followed to determine stair length is to lay out to scale the total rise and run on a steel square as shown in Fig. 146. Measuring between the two points on the legs of the square will give the stair length. While the stair length is not precisely the hypotenuse

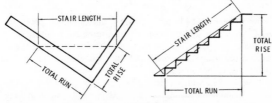

Fig. 146

of a right triangle whose sides are the total rise and the total run, in practice it is considered to be. The error as shown below is slight. Stair length may also be calculated using either trigonometry or the sum-of-the-squares equation.

The length of the rough stock needed for layout will depend on the style of construction. In many cases it must be greater than the stair length.

Stair Horse Layout

The stair horse should be laid out with the top edge of the stock as the layout *top line*. The edge of the steps are laid out touching this top line or layout line. This will result in a minimum amount of cutout and will leave supporting material below. The layout is started from the left end holding the body of the square in the left hand. The tongue is held in the right hand with the outside point of the square down or facing away from the top line, as shown in Fig. 147.

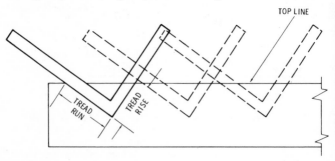

Fig. 147

As shown in Fig. 146, the tread-rise measurement point on the tongue of the square, and the tread-run measurement point on the body of the square, are positioned on the top line. The mark for the tread-run cut is made against the body and the tread-rise cut mark is made against the tongue of the square.

The square is then moved to the next position, again with the run and rise measurement points on the body and tongue matching the top line. The intersecting points of adjacent layouts become the points or edges of the steps. The square is moved along in this manner and layouts made until the required number of steps have been layed out.

Stair Stringer Layout

Because stair stringers usually act as enclosures on the sides, it is desirable to have stringer material extend above the treads. To provide this material above the treads a stair stringer is usually laid out from the bottom or *lower line*. This layout is also made starting from the left but the square is reversed as shown in Fig. 148.

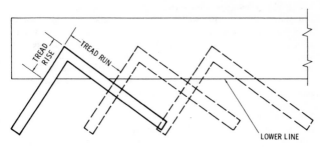

Fig. 148

Stair Drop

Layout procedure up to this point has made no provision for the thickness of the stair tread. To compensate for the tread thickness the stair horse or stringer must be "dropped." If this is not done the first step will be the thickness of the tread greater than the proper tread rise, and the top step will be

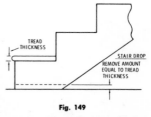

Fig. 149

the same amount less than the proper rise. Removing the correct amount from the bottom of the horse or stringer automatically corrects the height of both the bottom and top rises.

If finish flooring is to be laid after stairs are installed the amount that should be removed to allow for "drop" will be the difference between tread thickness and the finish flooring.

RIGGING

Weight Estimating

The first, and usually the most important consideration when selecting tools and equipment for rigging work is the weight of the object to be moved. Reasonable accuracy in the determination of an object's weight is a requirement for safe rigging. When it is not known, and dependable information is not available, the weight must be estimated. This should be an approximate calculation, not a guess.

In most cases a rigging weight estimate is made by roughly calculating the object's volume and multiplying this by the unit weight of the material of which it is made. As only an approximate figure is required, the calculation can be simplified by using approximate values which will allow many of the calculations to be made mentally.

For example, most heavy objects are made of iron or steel which ranges in weight from 475 to 490 pounds per cubic foot. This value can be rounded off to 500. For cylindrical calculations the value of Pi (π) can be rounded off from 3.1416 to an even 3. The object's dimensions can be rounded off to the closest even numbers, preferably multiples of 10. When rounding off dimensions, alternately increasing and decreasing to get even numbers will help to cancel out errors.

In the example below values and dimensions are rounded off to allow a quick and accurate weight estimate of the tank.

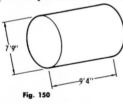

Fig. 150

¾″ *Thickness*

1 sq. ft. area has
1⁄16 cu. ft. volume
312 divided by 16 gives
20 cu. ft. *volume*
20 × 500 = 10,000 lbs. *total weight*

¾″ Thick
Steel Plate

Round off 7′9″ to 8′ dia. or 4′ rad.
Round off 9′4″ to 9′ length
Estimate Calculation
rad. × rad. × Pi. = end area
4 × 4 × 3 =   48 sq. ft. end area
dia. × Pi × length = shell area
8 × 3 × 9 = 216 sq. ft. shell area
                312 sq. ft. *total area*

An alternate method is to obtain the weight per square foot of steel plate from a table of weights, and multiply this by the total area. The weight per square foot of ¾-inch steel plate is listed as 30.6 pounds. Rounding off the values:

$$31 \times 310 = 9,600 \text{ lbs. } \textit{total weight}$$

Estimating the weight of any regular shaped object may be done in the same manner as the preceding example. In some cases several calculations may be required, as with the chambered roll illustrated in Fig. 151.

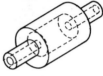

Fig. 151

For calculating purposes the roll is considered as made up of three parts, two shafts and a body. The total solid volume is the sum of the solid volumes of the three parts. To determine actual volume the total chamber volume, which is the sum of the three chamber volumes, is subtracted from the total solid volume. Weight is then determined by multiplying the actual volume by the unit weight of the roll material.

The weight of irregular shaped objects may be estimated with a high degree of accuracy by visualizing the object as a regular shape, or made up of a group of regular shapes. For example, the irregular shaped object in Fig. 152A may be visualized as a regular shaped object of lesser dimensions as shown in Fig. 152B.

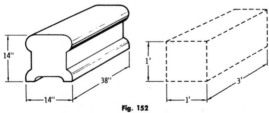

Fig. 152

Machines are usually an assembly of components of varying shapes, sizes and construction. They may be visualized as a group of regular shaped solid units when weight estimating. Each unit must be reduced in size to approximate the actual volume of material it contains. The machine shown in Fig. 153A for example could be visualized as shown in Fig. 153B.

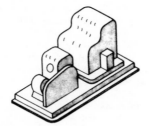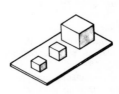

Fig. 153

Weight of Steel Bar Stock in Pounds per Lin. Foot

Size	Square	Round
1	3.4	2.7
1½	7.7	6.0
2	13.6	13.6
3	30.6	24.0
4	54.4	42.7
5	85.0	66.8
6	122.4	96.1
7	166.7	130.8
8	217.6	171.0
9	283.1	222.3
10	340.0	267.0
11	411.4	323.1
12	489.6	384.5

Weight of Steel Plate in lbs. per Sq. Foot

Thickness	Weight
1/16	2.55
1/8	5.1
3/16	7.65
1/4	10.2
5/16	12.75
3/8	15.3
1/2	20.4
5/8	25.5
3/4	30.6
1	40.8
1¼	51.0
1½	61.2
2	81.6

Weights of Materials

Material	Weight per cu. in.	Weight per cu. ft.
Aluminum	.093	160
Brass	.303	524
Cast Iron	.260	450
Concrete	.083	144
Sand	.070	120
Steel	.281	490
Water	.036	62½
Wood	.020	36

Wire Rope

The basic element in the construction of wire rope is a single metallic "wire." Several of these wires are laid helically around a *center* to form a *strand*. Finally, a number of strands are laid helically around a *core* to form the wire rope.

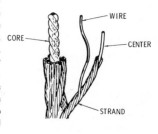

The primary function of the core is to serve as a foundation for the rope, to keep it round and to keep the strands correctly spaced and supported.

During construction the wires that

Fig. 154

make up the strand may be laid around the center in either a clockwise or counterclockwise direction. The same is true of the strands when they are laid around the core. This direction of rotation is called the *lay* of the rope. In "right" lay rope the strands rotate around the core in a clockwise direction, as the threads do in a right-hand thread. In "left" lay the strands rotate counterclockwise as do lefthand threads.

The terms *regular* and *lang* are used to designate direction of the wires around the center. "Regular" lay means that the wires rotate in a direction opposite to the direction of the strands around the core. This results in the wires being roughly parallel to the center line of the rope. "Lang" lay means the wires rotate in the same direction as the strands resulting in the wires being at a diagonal to the rope center line.

A right regular lay rope is shown in Fig. 154. The strands rotate clockwise and the wires counterclockwise. This is the most widely used rope lay and is commonly referred to simply as "regular lay."

Wire rope is classified by number of strands and the approximate number of wires in each strand. For example, the 6×7 classification indicates the rope has 6 strands and that each strand contains 7 wires. The wires in a strand are placed in layers around a center wire, each layer containing six more wires than the preceding one. These arrangements are referred to by the number of wires in each layer. The 7 wire strand is 6-1, the 19 wire strand is 12-6-1, the 37 wire strand is 18-12-6-1.

The designation for wire rope classifications are only nominal as the actual number of wires in a strand varies with the style of construction. For example, the 6 × 19 classification is made up of wire rope having anywhere from 15 to 26 wires per strand.

Factor of Safety

The safe use of wire rope requires that loads be limited to a portion of the rope's ultimate or breaking strength. The safe load for a wire rope is determined by dividing its breaking strength by a "factor of safety." Factors of safety for wire rope range from 5 for steady loads to 8 or more for uneven and shock loads.

For example, the breaking strength for a ½″ diameter improved-plow-steel rope is listed in Table 18 at 10.5 tons. If this rope were to be used with hoisting tackle at a factor of safety of 5, its maximum safe load would be ⅕th of the breaking strength or 2.1 tons. If however it were to be used in a sling at a factor of safety of 8, its maximum safe load would be ⅛th of the breaking strength or 1.4 tons.

Table 18. 6 x 19 Classification Ropes—Independent Wire Rope Core (IWRC)

Diameter in Inches	Breaking Strength in Tons of 2000 Pounds		
	Improved Plow Steel		Extra Improved Plow Steel
	Fiber Core	IWRC	IWRC
³⁄₁₆	1.46	—	—
¼	2.59	2.78	3.20
⁵⁄₁₆	4.03	4.33	4.98
⅜	5.77	6.20	7.14
⁷⁄₁₆	7.82	8.41	9.67
½	10.2	11.0	12.6
⁹⁄₁₆	12.9	13.9	15.9
⅝	15.8	17.0	19.6
¾	22.6	24.3	27.9
⅞	30.6	32.9	37.8
1	39.8	42.8	49.1

Table 19. Efficiencies of End Attachments

Fitting	Nominal Efficiency, percent of catalog rated rope strength
Wire Rope Sockets	100
Spelter (Zinced) Attachments	100
Fittings (Swaged or Pressed)	100
*Torpedo Collar (with or without thimble)	100
Open Wedge Sockets	80-90
Clips (U-bolt Type)	80
Clips (Twin-base Type)	80
Spliced-in Thimbles: ¼ and smaller	90
⁵⁄₁₆	89
⅜	88
½	86
⅝	84
¾	82
⅞ to 2½, incl.	80

6X7

6X19

6X37

Fig. 155

Wire Rope Attachment

The *U-bolt* or *Crosby* (Fig. 157A) wire rope clip is probably the most common method of attaching a wire rope to equipment. All U-bolt clips must be placed on the rope with U bolts bearing on the short or "dead" end of the rope. Illustrations of correct and wrong application of U-bolt clips are shown in Fig. 156.

An improved type wire rope clip called the *Double-base Safety* or *fist grip,* shown in Fig. 157B, has corrugated jaws to fit both parts of the rope, allowing it to be installed without regard to the live or dead part.

When making an eye attachment with clips, a thimble should be used and the correct number of clips (listed in Table 20). All clips should be spaced not less than six rope diameters apart. Apply the clip farthest from the thimble first, at about 4 inches from the end of the rope, and screw up tightly. Next, put on the clip nearest the thimble and apply the nuts handtight. Then put on the one or more intermediate clips handtight. Take a strain on the rope, and while the rope is under this strain, tighten all the clips previously left loose. Tighten alternately on the two nuts so as to keep the clip square. After the rope has been in use a short time retighten all clips.

CORRECT METHOD:
U-BOLTS OF CLIPS ON
SHORT END OF ROPE

WRONG:
U-BOLTS ON LIVE
END OF ROPE

WRONG:
STAGGERED CLIPS;
TWO CORRECT AND
ONE WRONG

CORRECT METHOD
(CLIPS REMOVED):
NO DISTORTION ON
LIVE END OF ROPE

WRONG:
NOTE MASHED SPOTS
ON LIVE END OF ROPE

WRONG:
NOTE MASHED SPOT
DUE TO U-BOLT OF
CENTER CLIP

Fig. 156

Table 20. Recommended Number of Clips

Rope Size	U-Bolt	Safety
¼ to ⅜	2 to 4	2
⁷⁄₁₆ to ⅝	3 or 4	2
¾	4 or 5	3
⅞ to 1⅛	4 or 5	4
1¼ to 1½	5 to 8	5

Wedge socket attachments are used on equipment where frequent changes are required. Care must be exercised to install the rope so that the pulling part is directly in line with the clevis pin as shown in Fig. 157C. If incorrectly installed, as shown in Fig. 157D, a sharp bend will be produced in the rope as it enters the socket.

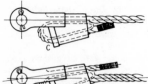

Fig. 157

Multiple Reeving

A single rope supporting a load is referred to as a single part line and the tension in the rope is equal to the suspended weight. When a load is supported by a multipart wire rope tackle as shown in Fig. 158, and the rope is not moving, the load on each line including the lead line is equal to the weight of the load divided by the number of parts of rope supporting the load. When this load is raised, however, the loads on the individual supporting ropes change, increasing from the dead end to the lead line.

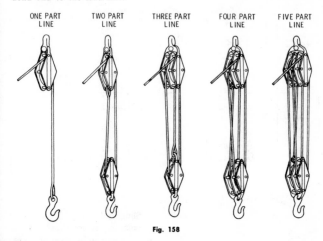

ONE PART   TWO PART   THREE PART   FOUR PART   FIVE PART
LINE   LINE   LINE   LINE   LINE

Fig. 158

How to Measure Wire Rope

The measurement should be made carefully with calipers. Fig. 159 shows the correct and incorrect method of measuring the diameter of wire rope.

Wire Rope Slings

The single sling with loop ends, most widely used of all sling types, lends itself readily for use in a basket hitch, choker hitch, or as a straight rope. Blocking should always be used to protect sling from sharp corners.

TRUE DIAMETER

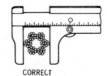

CORRECT

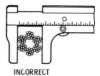

INCORRECT

Fig. 159

Table 21. Safe Loads in Tons

Nominal Size Inches	Single	Choker	U-sling	Basket	60-deg.	45-deg.	30-deg.
¼	.5	.3	.7	.6	.57	.5	.3
⁵⁄₁₆	.8	.6	1.1	1.0	.9	.7	.6
⅜	1.1	.8	1.5	1.4	1.3	1.1	.8
½	2.0	1.4	2.7	2.4	2.3	1.9	1.3
⅝	2.9	2.1	4.2	3.8	3.7	3.0	2.1
¾	4.1	3.0	6.0	5.4	5.2	4.2	3.0
⅞	5.6	3.8	7.7	6.8	6.7	5.4	3.8
1	7.2	5.0	10.0	9.3	8.7	7.1	5.0
1⅛	9.0	5.6	11.2	10.5	9.7	7.9	5.6

LOOP

THIMBLE

HOOK AND THIMBLE

SHACKLE & THIMBLE

LINK AND THIMBLE

TWIN THIMBLE

CLOSED SOCKET

OPEN SOCKET

Fig. 160

Sling end fittings of the more popular styles are shown at the right. Any combination of these, to suit the job requirements, are available.

The lengths of slings having loop or ring style end fittings are measured from the weight bearing surface. Those with pin style end fittings are measured from from the center line of the pin.

Eyebolts and Shackles

The strength of eyebolts is influenced greatly by the direction of pull to which it is subjected. For loads involving angular forces the shoulder type has several times the strength of the conventional eyebolt. Table 22 lists the safe loads for the two types by sizes and direction of load.

The shackle is the recommended fastener for attaching slings to eyebolts, chain, or wire rope. Table 23 lists the safe loads by nominal size and gives the dimensions of standard shackles.

Table 22. Ordinary Drop-Forged Steel Eyebolts

Size			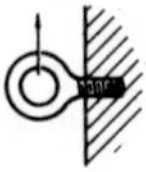
½"	1,100 lb.	50 lb.	40 lb.
⅝"	1,800 lb.	90 lb.	65 lb.
¾"	2,800 lb.	135 lb.	100 lb.
⅞"	3,900 lb.	210 lb.	150 lb.
1"	5,100 lb.	280 lb.	210 lb.
1¼"	8,400 lb.	500 lb.	370 lb.
1½"	12,200 lb.	770 lb.	575 lb.
1¾"	16,500 lb.	1,080 lb.	800 lb.
2"	21,800 lb.	1,440 lb.	1,140 lb.

Drop Forged Steel Shoulder-type Eyebolts

Size			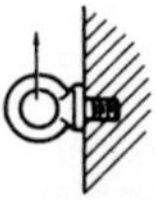
¼"	300 lb.	30 lb.	40 lb.
½"	1,300 lb.	140 lb.	150 lb.
¾"	3,000 lb.	250 lb.	300 lb.
1"	6,000 lb.	500 lb.	600 lb.
1¼"	9,000 lb.	800 lb.	900 lb.
1½"	13,000 lb.	1,200 lb.	1,300 lb.
2"	23,000 lb.	2,100 lb.	2,300 lb.
2½"	37,000 lb.	3,800 lb.	4,300 lb.

Table 23. Anchor Shackles

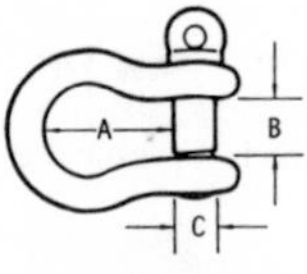

Nominal Size	Tons Safe Load	Dimensions		
		A	B	C
3/8	.8	1½	11/16	7/16
½	1.4	2	7/8	5/8
5/8	2.2	2⅜	1 1/16	¾
¾	3.2	2⅞	1¼	7/8
7/8	4.3	3¼	1⅜	1
1	5.6	3⅝	1 11/16	1⅛
1⅛	6.7	4¼	1⅞	1¼
1¼	8.2	4¾	2	1⅜
1½	11.8	5½	2¼	1⅝
2	21.1	7¾	3¼	2¼

Screw Pin—Drop
Forged Steel

Knots

Bight—Formed by simply bending the rope and keeping the sides parallel.

Loop or Turn—Formed by crossing the sides of a bight.

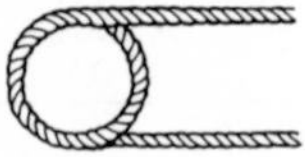

Round Turn—Further bending of one side of a loop.

Standing Part—That part of a rope that is not used in tying a knot; the long part which is not worked upon.

End—As the name implies, the very end of the rope.

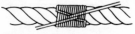

Whipping—Common, plain or ordinary whipping is tied by laying a loop along the rope and then making a series of turns over it. The working end is finally stuck through this loop and the end hauled back out of sight. Both ends are then trimmed short. A whipping should be, in width, about equal to the diameter of the rope.

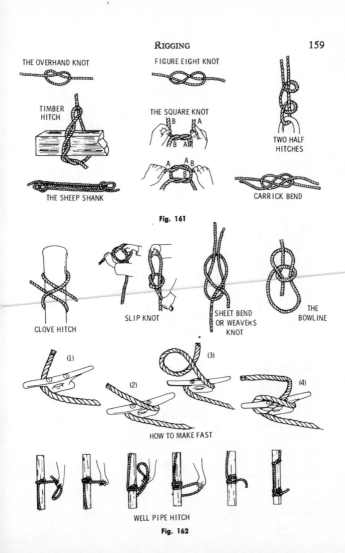

THE OVERHAND KNOT

FIGURE EIGHT KNOT

TIMBER HITCH

THE SQUARE KNOT

TWO HALF HITCHES

THE SHEEP SHANK

CARRICK BEND

Fig. 161

CLOVE HITCH

SLIP KNOT

SHEET BEND OR WEAVERS KNOT

THE BOWLINE

(1)

(2)

(3)

(4)

HOW TO MAKE FAST

WELL PIPE HITCH

Fig. 162

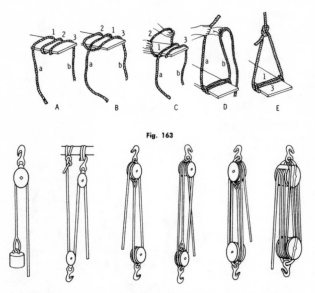

Fig. 163

Fig. 164

Lay the short end (a) of the rope over the top of the plank (Fig. 163A) leaving enough handing down to the left to tie to the long rope, as shown in Fig. 163E. Wrap the long end (b) loosely twice around the plank, letting it hang down to the right as shown in Fig. 163A. Now, carry rope 1 over rope 2 and place it next to rope 3 as shown in Fig. 163B. Pick up rope 2 (Fig. 163C) and carry it over 1 and 3, and over the end of the plank. Take up the slack by pulling rope (a) to the left and rope (b) to the right. Draw ropes (a) and (b) above the plank as shown in Fig. 163D and join the short end (a) to the long rope (b) by an overhand bowline as shown in Fig. 163E. Pull the bowline tight, at the same time adjusting the lengths of the two ropes so that they hold the plank level. Attach a second rope to the other end of the plank in the same way and the scaffold is now ready for safe use.

PIPING

Thread Designations

American Standard

American Standard Pipe Threads are designated by specifying in sequence the nominal size, number of threads per inch and the thread series symbols.

Nominal Size	No. of Threads	Symbols
3/8	18	NPT

Each of the letters in the symbols have the following significance.

N—American (Nat.) Standard   S—Straight
.P—Pipe   L—Locknut
T—Taper   R—Railing Fittings
C—Coupling   M—Mechanical

Examples

3/8—18 NPT	American Standard Taper Pipe Thread
3/8—18 NPSC	Am. Std. Straight Coupling Pipe Thread
1/8—27 NPTR	Am. Std. Taper Railing Pipe Thread
1/2—14 NPSM	Am. Std. Straight Mechanical Pipe Thread
1—11½ NPSL	Am. Std. Straight Locknut Pipe Thread

Left-hand threads are designated by adding LH

American Standard Taper Pipe Thread Form

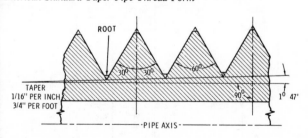

Fig. 165

American Standard Taper Pipe Threads (NPT)

Basic Dimensions

Taper pipe threads are engaged or made-up in two phases, *hand engagement* and *wrench makeup*. The table below lists the basic hand engagement and wrench makeup for American Standard Taper Pipe Threads. Dimensions are rounded off to the closest $\frac{1}{32}$ inch.

Note: Table 24 is a table of basic makeup dimensions. Commercial product may vary as much as one turn large or small and still be within standard tolerance. In actual shop practice pipe threads are usually cut to give a connection which makes up less than the basic standard. Common practice is about 3 turns by hand and 3 to 4 turns by wrench.

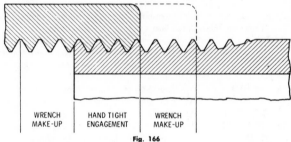

WRENCH MAKE-UP   HAND TIGHT ENGAGEMENT   WRENCH MAKE-UP

Fig. 166

Table 24. Taper Pipe Makeup Dimensions

Pipe Size	Threads Per. In.	HAND TIGHT		WRENCH MAKE-UP		Dimension	Turns
		Dimension	Turns	Dimension	Turns	TOTAL MAKE-UP	
$\frac{1}{8}$	27	$\frac{3}{16}$	$4\frac{1}{2}$	$\frac{3}{32}$	$2\frac{1}{2}$	$\frac{9}{32}$	7
$\frac{1}{4}$	18	$\frac{7}{32}$	4	$\frac{3}{16}$	3	$\frac{13}{16}$	7
$\frac{3}{8}$	18	$\frac{1}{4}$	$4\frac{1}{2}$	$\frac{3}{16}$	3	$\frac{7}{16}$	$7\frac{1}{2}$
$\frac{1}{2}$	14	$\frac{5}{16}$	$4\frac{1}{2}$	$\frac{7}{32}$	3	$\frac{17}{32}$	$7\frac{1}{2}$
$\frac{3}{4}$	14	$\frac{5}{16}$	$4\frac{1}{2}$	$\frac{7}{32}$	3	$\frac{17}{32}$	$7\frac{1}{2}$
1	$11\frac{1}{2}$	$\frac{3}{8}$	$4\frac{1}{2}$	$\frac{1}{4}$	$3\frac{1}{4}$	$\frac{5}{8}$	$7\frac{3}{4}$
$1\frac{1}{4}$	$11\frac{1}{2}$	$\frac{13}{32}$	$4\frac{1}{2}$	$\frac{9}{32}$	$3\frac{1}{4}$	$\frac{11}{16}$	8
$1\frac{1}{2}$	$11\frac{1}{2}$	$\frac{13}{32}$	$4\frac{1}{2}$	$\frac{9}{32}$	$3\frac{1}{4}$	$\frac{11}{16}$	8
2	$11\frac{1}{2}$	$\frac{7}{16}$	5	$\frac{1}{4}$	3	$\frac{11}{16}$	8
$2\frac{1}{2}$	8	$\frac{11}{16}$	$5\frac{1}{2}$	$\frac{3}{8}$	3	$1\frac{1}{16}$	$8\frac{1}{2}$
3	8	$\frac{3}{4}$	6	$\frac{3}{8}$	3	$1\frac{1}{8}$	9

Pipe Measurement

Dimensions on pipe drawings specify the location of center lines and/or points on center lines, they do not specify pipe lengths. This system of distance dimensioning and measurement is also followed in the fabrication and installation of pipe assemblies.

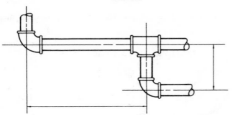

Fig. 167

To determine actual pipe lengths, allowances must be made for the length of the fittings and the distance threaded pipe is made-up into the fittings. The method of doing this is to subtract an amount called *take-out* from the *center-to-center* dimension. The relationship of take-out to other threaded pipe connection distances termed, *make-up, center-to-center,* and *end-to-end* are illustrated in Fig. 168.

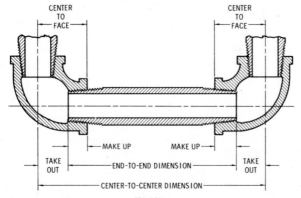

Fig. 168

To determine end-to-end pipe length the take-out is subtracted from the center-to-center dimension. Standard tables may be used for this purpose. These tables should be used with judgment however, since commercial product tolerance is one turn plus or minus. On critical connections, materials should be checked and compensation made for variances.

Table 25. Takeout Allowances

Pipe Size	90° Elbow		Tee		45° Elbow	
	A	Take Out	B	Take Out	C	Take Out
1/8	11/16	7/16	11/16	7/16	9/16	1/4
1/4	13/16	7/16	13/16	7/16	3/4	3/8
3/8	15/16	9/16	15/16	9/16	13/16	7/16
1/2	1 1/8	5/8	1 1/8	5/8	7/8	3/8
3/4	1 5/16	3/4	1 5/16	3/4	1	7/16
1	1 1/2	7/8	1 1/2	7/8	1 1/8	9/16
1 1/4	1 3/4	1 1/8	1 3/4	1 1/8	1 5/16	11/16
1 1/2	1 15/16	1 1/4	1 15/16	1 1/4	1 7/16	3/4
2	2 1/4	1 5/8	2 1/4	1 5/8	1 11/16	1

Pipe Size	Thread Make-Up	Coupling		Union	
		D	Take Out	E	Take Out
1/8	1/4	1	1/4	1 1/2	3/4
1/4	3/8	1 1/8	3/8	1 5/8	7/8
3/8	3/8	1 1/4	3/8	1 3/4	1
1/2	1/2	1 3/8	3/8	1 7/8	1
3/4	9/16	1 1/2	3/8	2 1/8	1 1/16
1	9/16	1 3/4	1/2	2 3/8	1 1/4
1 1/4	5/8	2	3/4	2 5/8	1 3/8
1 1/2	5/8	2 1/8	7/8	3	1 1/2
2	11/16	2 1/2	1 1/4	3 1/4	1 3/4

As the threads are made up in the fitting, high forces and pressures are developed by the wedging action of the taper. Also, frictional heat is developed as the surfaces are deformed to match the variations in the thread form. It is important that the threads be clean and well lubricated, and that the connection is not screwed up fast enough to generate excessive heat. Use of a lubricant called "dope" allows thread surfaces to deform and mate without galling and seizing. The dope also helps to plug openings resulting from improper threads and acts as a cement.

The requirements for tight makeup of threaded pipe connections are: good quality threads; clean threads; proper dope for the application; slow final makeup to avoid heat generation.

Table 26. Commercial Pipe Sizes and Wall Thicknesses

Nominal Pipe Size	Outside Dia.	Nominal Wall Thickness					
		Sched. 5	Sched. 10	Sched. 40 Std.	Sched. 80 Ex. St.	Sched. 160	Ex. Ex. Strong
1/8	.405	—	.049	.068	.095	—	—
1/4	.540	—	.065	.088	.119	—	—
3/8	.675	—	.065	.091	.126	—	—
1/2	.840	—	.083	.109	.147	.187	.294
3/4	1.050	.065	.083	.113	.154	.218	.308
1	1.315	.065	.109	.133	.179	.250	.358
1 1/4	1.660	.065	.109	.140	.191	.250	.382
1 1/2	1.900	.065	.109	.145	.200	.281	.400
2	2.375	.065	.109	.154	.218	.343	.436
2 1/2	2.875	.083	.120	.203	.276	.375	.552
3	3.500	.083	.120	.216	.300	.438	.600
3 1/2	4.000	.083	.120	.226	.318	—	—
4	4.500	.083	.120	.237	.337	.531	.674
5	5.563	.109	.134	.258	.375	.625	.750
6	6.625	.109	.134	.280	.432	.718	.864
8	8.625	.109	.148	.322	.500	.906	.875

Pipe schedule numbers indicate pipe strength. The higher the number in a given size, the greater the strength. The schedule indicates the approximate values of the expression:

$$\text{Schedule Number} = \frac{1000 \times \text{internal pressure}}{\text{allowable stress in pipe}}$$

Copper Water Tube

Seamless copper water tube, popularly called *copper tubing,* is widely used for plumbing, water lines, heater coils, fuel oil lines, gas lines, etc. Standard copper tubing is commercially available in three (3) types designated as type "K", "L", and "M". The type "K" has the thickest wall and is generally used for underground installations. The type "L" has a thinner wall and is most widely used for general plumbing, heating, etc. The type "M" is the thin-wall tubing used for low-pressure and drainage applications.

The outside diameter of copper tubing is uniformly ⅛″ larger than the nominal size, for example the outside diameter of 1″ tubing is 1⅛″. All three types of copper tubing in any given size have the same outside diameter. The inside diameter of each type will vary as the wall thickness varies.

Table 27. Standard Copper Tubing Dimensions

Nominal Size	Outside Dia.	Inside Diameter		
		Type "K"	Type "L"	Type "M"
⅜	½	.402	.430	.450
½	⅝	.528	.545	.569
⅝	¾	.652	.668	.690
¾	⅞	.745	.785	.811
1	1⅛	.995	1.025	1.055
1¼	1⅜	1.245	1.265	1.291
1½	1⅝	1.481	1.505	1.527
2	2⅛	1.959	1.985	2.009
2½	2⅝	2.435	2.465	2.495
3	3⅛	2.907	2.945	2.981
3½	3⅝	3.385	3.425	3.459
4	4⅛	3.857	3.905	3.935

The three types of tubing are made in either "hard" or "soft" grade. Hard .tubing is used for applications where lines must be straight without kinks or pockets. Soft tubing is used for bending around obstructions, inaccessible places etc.

45-Degree Offset

The *travel* distance of a 45-degree offset is calculated in the same manner as the diagonal of a square. Multiply the distance across flats by 1.414. The "run" and "offset" represent the two equal sides of a square and the "travel" the diagonal, as shown in Fig. 169.

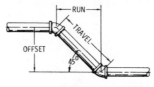

Fig. 169

There are also occasions when the travel is known and the offset and run dimensions are wanted. This may be done in the same manner as finding the distance across the flats of a square when the across corners dimension is known. Multiply the travel dimension by 0.707.

Examples

What is the "travel of a 16 inch 45 degree offset?

16 × 1.414 = 22.625 or 22⅝ inches.

What is the "offset" and "run" of a 45 degree offset having a travel of 26 inches?

26 × 0.707 = 18.382 or 18⅜ inches.

Other Offsets

The dimensions of piping offsets of several other common angles may be calculated by multiplying the known values by the appropriate constants listed in the following table:

Angle	To find Travel Offset Known	To Find Travel Run Known	To Find Run Travel Known	To Find Offset Travel Known
60°	1.155	2.000	0.500	0.866
30°	2.000	1.155	0.866	0.500
22½°	2.613	1.082	0.924	0.383
11¼°	5.126	1.000	0.980	0.195

45-Degree Rolling Offset

The 45-degree offset is often used to offset a pipe line in a plane other than the horizontal or vertical. This is done by rotating the offset out of the horizontal or vertical plane and is known as a *rolling offset*. The rolling offset can best be visualized as contained in an imaginary isometric box as shown in Fig. 170.

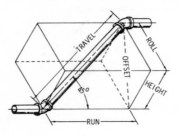

Fig. 170

The run and offset distances are equal, as they are in the plain 45-degree offset, however, there are two additional dimensions, *roll* and *height*. Two right-angle triangles must now be considered, the original one remains the same, with the offset and run as equal sides and the travel as the hypotenuse. The new triangle has the roll and height as sides and the offset as the hypotenuse.

The method of finding distances for plain 45-degree offsets is also used for calculating rolling offset distances. In addition the sum of the squares equations are used to find the values of the second triangle.

Sum Of The Squares—The sum-of-the-squares equation states that the hypotenuse of a right angle triangle squared, is equal to the side opposite squared plus the side adjacent squared. This equation is commonly written as follows:

$$c^2 = a^2 + b^2$$

Substitution of pipe offset terms in this equation and its rearrangements gives the following equations:

$$offset^2 = roll^2 + height^2$$
$$run^2 = roll^2 + height^2$$
$$roll^2 = offset^2 - height^2$$
$$roll^2 = run^2 - height^2$$
$$height^2 = offset^2 - roll^2$$
$$height^2 = run^2 - roll^2$$

Depending on what the known values are, it may sometimes be necessary to solve two equations to find the distance wanted.

Examples

What is the *travel* for a 6″ *roll* with a 7″ *height?*

Using the formula:

$$offset^2 = roll^2 + height^2$$

$offset^2 = (6 \times 6) + (7 \times 7) = 36 + 49 = 85$

$offset = \sqrt{85} = 9.22 = 9\frac{7}{32}$ inches

Followed by:

$$travel = offset \times 1.414$$

$travel = 9\frac{7}{32} \times 1.414 = 13.035 = 13\frac{1}{32}$ inches

What is the *roll* for a 11″ *offset* with 8″ *height?*

Using the formula:

$$roll^2 = offset^2 - height^2$$

$roll^2 = (11 \times 11) - (8 \times 8) = 121 - 64 = 57$

$roll = \sqrt{57} = 7.55 = 7\frac{9}{16}$ inches

What is the *height* for a 16″ *offset* with 12″ *roll?*

Using the formula:

$$height^2 = offset^2 - roll^2$$

$height^2 = (16 \times 16) - (12 \times 12) = 256 - 144 = 112$

$height = \sqrt{112} = 10.583 = 10\frac{9}{32}$ inches

Flanged Pipe Connections

Flanged pipe connections are widely used, particularly on larger size pipe, as they provide a practical and economical piping connection system. Flanges are commonly connected to the pipe by screw threads or by welding. Several types of flange facings are in use, the simplest of which are the plain *flat face,* and the *raised face.*

The plain flat-faced flange is usually used for cast iron flanges where pressures are under 125 pounds. Higher pressure cast iron flanges and steel flanges are made with a raised face. Generally full-face gaskets are used with flat-face flanges and ring gaskets with raised-face flanges. The function of the gasket is to provide a loose compressible substance between the faces with sufficient body resiliency and strength to make the flange connection leak proof.

The assembly and tightening of a pipe-flange connection is a relatively simple operation, however certain practices must be followed to obtain a leakproof connection. The gasket must line up evenly with the inside bore of the flange face with no portion of it extending into the bore. When tightening the bolts the flange faces must be kept parallel and the bolts tightened uniformly.

The tightening sequence for round flanges is shown in Fig. 171A. The sequence is to lightly tighten the first bolt then move directly across the circle for the second bolt, then move ¼ way around the circle for the third and directly across for the fourth, continuing the sequence until all are tightened.

When tightening an oval flange the bolts are tightened across the short center line first and then alternately from side to side moving away from the short center line as shown in Fig. 171B.

A four-bolt flange, either round or square, is tightened with a simple criss-cross sequence as shown in Fig. 171C.

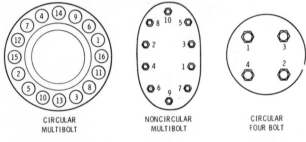

CIRCULAR MULTIBOLT NONCIRCULAR MULTIBOLT CIRCULAR FOUR BOLT

Fig. 171

Do not snug up bolts on first go around. This can tilt flanges out of parallel. If using an impact wrench, set the wrench at about ½ final torque for first go around. Pay particular attention to the hard-to-reach bolts.

Pipe-Flange Bolt-Hole Layout

The mating of pipe flanges to other flanges or circular parts requires correct layout of bolt holes. In addition, the holes must be located around the circle to line-up when the flanges are mated. The usual practice is to specify the location of the holes as either "On" or "Off" the vertical center line. The shop term commonly applied to "On" the center line layout is "One Hole Up" and to "Off" the center line is "Two Holes Up". An "Off" the center line or "Two Holes Up" layout is illustrated below.

While bolt holes may be layed out with a protractor using angular measurements to obtain uniform spacing, this method is most satisfactory when there are six (6) or less holes. Also, layout by stepping off spaces around the circle with dividers by trial and error, is a time consuming operation. To eliminate the trials and errors, a system of multipliers or constants may be used to calculate the chordal distance between bolt hole centers. Simply multiply the constant for the appropriate number of bolt holes by the bolt circle diameter to determine the chordal distance between holes.

Layout Procdure

1. Lay out horizontal and vertical center lines.
2. Lay out bolt circle.
3. Find value of "B" (multiply bolt circle dia. by constant).
4. For two-holes-up layout, divide "B" by 2 for value of "C."
5. Measure distance "C" off the center line and locate the center of the first bolt hole.
6. Set dividers to dimension "B" and layout center points by swinging arcs starting from first center point.

Table 28. Flange Hole Constants

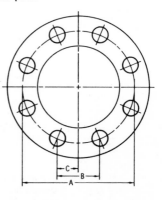

Number Of Bolt Holes	Constant
4	.707
6	.500
8	.383
10	.309
12	.259
16	.195
20	.156
24	.131
28	.112
32	.098
36	.087
40	.079

Valves

The principal function of two-way pipe valves is to open and close a line to flow. Commonly used for this purpose are the *gate, globe, needle,* and *plug* valves.

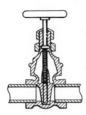

Gate—Fluid flows through a gate valve in a straight line. Its construction offers little resistance to flow and causes a minimum of pressure drop. A gate-like disc—actuated by a stem screw and handwheel—moves up and down at right angles to the path of flow, and seats against two seat faces to shut off flow.

Gate valves are best for services that require infrequent valve operation and where disc is kept either fully opened or closed.

Fig. 172

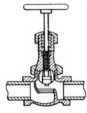

Globe—Fluid changes direction when flowing through a globe valve. The construction increases resistance to—and permits close regulation of—fluid flow. Disc and seat can be quickly and conveniently replaced or reseated. Angle valves are similar in design to globe valves. They are used when making a 90-degree turn in a line as they reduce the number of joints and give less resistance to flow than the elbow and globe valve.

Fig. 173

Needle—The needle valve can be used to shutoff flow, however, it is designed primarily as a throttling valve. The pointed disc can be adjusted in the mating seat to give small increments of flow change. As the port diameter is smaller than the connection size, resistance to flow is high, making the needle valve unsuitable for high-volume flow.

Fig. 174

Plug—Fluid flows through a plug valve in a straight line. Its advantages are low cost, small pressure loss because of its straight-through construction, and its fast operation. Only a quarter turn is needed to fully open or close it.

Fig. 175

Steam Traps

To efficiently drain condensate from steam lines, steam traps must be correctly located and the piping properly arranged. The following basic rules will go a long way toward providing satisfactory operation of steam traps (Fig. 176).

1. Provide a separate trap for each piece of equipment or apparatus. Short circuiting (steam follows path of least resistance to trap) may occur if more than one piece of apparatus, coil, etc., is connected to a single trap.

2. Tap steam supply off the top of the steam main to obtain dry steam and avoid steam line condensate.

3. Install a supply valve close to the steam main to allow maintenance and/or revisions without steam main shutdown.

4. Install a steam supply valve close to the equipment entrance to allow equipment maintenance work without supply line shutdown.

5. Connect condensate discharge line to lowest point in equipment to avoid water pockets and water hammer.

6. Install shutoff valve upstream of condensate-removal piping to cut off discharge of condensate from equipment and allow service work to be performed.

7. Install strainer and strainer flush valve ahead of trap to keep rust, dirt, and scale out of working parts and to allow blow-down removal of foreign material from strainer basket.

8. Provide unions on both sides of trap for its removal and/or replacement.

9. Install test valve downstream of trap to allow observance of discharge when testing.

10. Install check valve downstream of trap to prevent condensate flow-back during shutdown or in the event of unusual conditions.

11. Install downstream shutoff valve to cut off equipment condensate piping from main condensate system for maintenance or service work.

12. Do not install a bypass unless there is some urgent need for it. Bypasses are an additional expense to install and are frequently left open, resulting in loss of steam and inefficient operation of equipment.

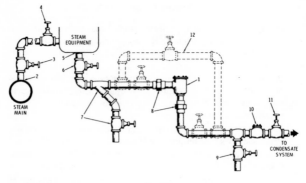

Fig. 176

Steam Trap Testing

In cases of improper functioning of steam equipment a few simple checks of the steam system should be made before looking for trap malfunction. The following preliminary checks should precede checking the operation of a steam trap.

1. Check the steam supply pressure—it should be at or above the minimum required.
2. Check to be sure all valves required to be open, are in the full-open position (supply, upstream shutoff, downstream shutoff).
3. Check to be sure all valves required to be closed, are in the tight closed position (bypass, strainer, test).

The initial step in checking the operation of a steam trap is to check its temperature. Since a properly functioning steam trap is an automatic valve which allows condensate to be discharged but closes to prevent the escape of steam, it should operate very close to the steam temperature. For exact checks a surface pyrometer should be used. A convenient and dependable operating test is simply to sprinkle water on the trap. If the water spatters, rapidly boils, and vaporizes, the trap is hot and probably is very close to steam temperature.

If it is found that the trap is very close to steam temperature the next step is to determine if condensate or steam is being discharged. When a test valve is provided for this purpose, the check is made by closing the downstream shutoff valve and opening the test valve. The discharge from the test valve should be carefully observed to determine if condensate or live steam is escaping. If the trap being tested is the type that has an opening and closing cycle, condensate should flow from the test valve and then it should stop as the trap shuts off. The flow should resume when the trap opens again, etc. Steam should not discharge from the test valve if the trap is operating properly. If the trap is a continuous discharge type, there should be a continuous discharge of condensate, but no steam.

In the event the installation does not have a test valve, the trap may be checked by listening to its operation. The ideal instrument to do this is an industrial stethoscope. If not available, a suitable device for this purpose is a screw driver or metal rod. By holding one end against the trap and the other end against the ear the sound of the traps operation may be heard. If the trap is operating properly the flow of condensate should be heard for a few seconds, a click as the valve closes, and then silence, indicating the valve has closed tight. This cycle of sounds should repeat in a regular pattern. The listening procedure however is not suitable for checking continuous discharge traps as these automatically regulate to an open position in balance with the condensate flow.

Another check on trap operation is to open the strainer valve and observe the discharge at this point. There should be an initial gush of condensate and steam as the valve is opened, followed by a continuous flow of live steam. If condensate flows for a prolonged period before steam is observed, the condensate is not being properly discharged from the system.

Unsatisfactory performance of a steam unit may not be due to improper steam trap operation. When testing steam traps conditions other than trap malfunction must also be considered. Some of the common faults that cause troubles are the following.

1. Inadequate steam supply.
2. Incorrectly sized trap.
3. Improperly connected piping.
4. Improper pitch of condensate lines.
5. Inadequate condensate lines.

Table 29. Steam Systems Troubleshooting Guide

CONDITION	REASON	CORRECTIVE ACTION

TRAP BLOWS LIVE STEAM	1. NO PRIME (bucket traps)
	  a. Trap not primed when originally installed
	  b. Trap not primed after cleanout
	  c. Open or leaking by-pass valve
	  d. Sudden pressure drops
		1.
		  a. Prime the trap
		  b. Prime the trap
		  c. Remove or repair by-pass valve
		  d. Install check valve ahead of trap

	2. VALVE MECHANISM DOES NOT CLOSE
	  a. Scale or dirt lodged in orifice
	  b. Worn or defective valve or disc mechanism
		2.
		  a. Clean out the trap
		  b. Repair or replace defective parts

	3. RUPTURED BELLOWS (thermostatic traps)	3. Replace bellows

	4. BACK PRESSURE TOO HIGH (thermodynamic trap)
	  a. Worn or defective parts
	  b. Trap stuck open
	  c. Condensate return line or pig tank undersize
		4.
		  a. Repair or replace defective parts
		  b. Clean out the trap
		  c. Increase line or pig tank size

	5. BLOWING FLASH STEAM Forms when condensate released to lower or atmospheric pressure	5. Normal condition No corrective action necessary

Table 29. Steam Systems Troubleshooting Guide (Cont.)

CONDITION	REASON	CORRECTIVE ACTION
TRAP DOES NOT DISCHARGE	1. PRESSURE TOO HIGH a. Trap pressure rating too low b. Orifice enlarged by normal wear c. Pressure reducing valve set too high or broken d. System pressure raised	1. a. Install correct trap b. Replace worn orifice c. Readjust or replace pressure reducing valve d. Install correct pressure change assembly
	2. CONDENSATE NOT REACHING TRAP a. Strainer clogged b. Obstruction in line to trap inlet c. By-pass opening or leaking d. Steam supply shut off	2. a. Blow out screen or replace b. Remove obstruction c. Remove or repair by-pass valve d. Open steam supply valve
	3. TRAP CLOGGED WITH FOREIGN MATTER	3. Clean out and install strainer
	4. TRAP HELD CLOSED BY DEFECTIVE MECHANISM	4. Repair or replace mechanism
	5. HIGH VACUUM IN CONDENSATE RETURN LINE	5. Install correct pressure change assembly
	6. NO PRESSURE DIFFERENTIAL ACROSS TRAP a. Blocked or restricted condensate return line b. Incorrect pressure change assembly	6. a. Remove restriction. b. Install correct pressure change assembly

Table 29. Steam Systems Troubleshooting Guide (Cont.)

CONDITION	REASON	CORRECTIVE ACTION
	1. TRAP TOO SMALL	1.
	a. Capacity undersized	a. Install properly sized larger trap
	b. Pressure rating of trap too high	b. Install correct pressure change assembly
	2. TRAP CLOGGED WITH FOREIGN MATTER	2.
	a. Dirt or foreign matter in trap internals	a. Clean out and install strainer
	b. Strainer plugged	b. Clean out strainer
CONTINUOUS DISCHARGE FROM TRAP	3. BELLOWS OVERSTRESSED (Thermostatic traps)	3. Replace bellows
	4. LOSS OF PRIME	4. Install check valve on inlet side
	5. FAILURE OF VALVE TO SEAT	5.
	a. Worn valve and seat	a. Replace worn parts
	b. Scale or dirt under valve and in orifice	b. Clean out the trap
	c. Worn guide pins and lever	c. Replace worn parts

Table 29. Steam Systems Troubleshooting Guide (Cont.)

CONDITION	REASON	CORRECTIVE ACTION
SLUGGISH OR UNEVEN HEATING	1. TRAP HAS NO CAPACITY MARGIN FOR HEAVY STARTING LOADS	1. Install properly sized larger trap
	2. INSUFFICIENT AIR HANDLING CAPACITY (bucket traps)	2. Use thermic buckets or increase vent size
	3. SHORT CIRCUITING (group traps)	3. Trap each unit individually
	4. INADEQUATE STEAM SUPPLY a. Steam supply pressure valve has changed b. Pressure reducing valve setting off	4. a. Restore normal steam pressure b. Readjust or replace reducing valve
BACK PRESSURE TROUBLES	1. CONDENSATE RETURN LINE TOO SMALL	1. Install larger condensate return line
	2. OTHER TRAPS BLOWING STEAM INTO HEADER	2. Locate and repair other faulty traps
	3. PIG TANK VENT LINE PLUGGED	3. Clean out pig tank vent line
	4. OBSTRUCTION IN CONDENSATE RETURN LINE	4. Remove obstruction
	5. EXCESS VACUUM IN CONDENSATE RETURN LINE	5. Install correct pressure change assembly

STEAM LINES

Opening Steam Supply Valve

The opening of valves controlling steam flow in steam supply lines, called steam *mains,* requires care and correct procedure. The expansion or growth of the piping system as the temperature increases when steam is introduced, must be carefully controlled. Also the air in the line and the large volume of condensate formed as the line heats up, must be removed. To facilitate removal of condensate during normal operation, as well as at start-up, steam lines incorporate *drip pockets, drip legs, drip valves* as shown in Fig. 177.

The following procedure should be followed when opening a steam main supply valve (See Fig. 177.)

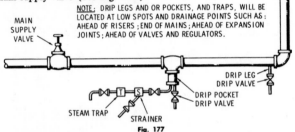

NOTE: DRIP LEGS AND OR POCKETS, AND TRAPS, WILL BE LOCATED AT LOW SPOTS AND DRAINAGE POINTS SUCH AS: AHEAD OF RISERS; END OF MAINS; AHEAD OF EXPANSION JOINTS; AHEAD OF VALVES AND REGULATORS.

MAIN SUPPLY VALVE

DRIP LEG
DRIP VALVE

DRIP POCKET
DRIP VALVE

STEAM TRAP   STRAINER

Fig. 177

1. Open all drip valves full open to act as air vents and condensate discharge openings. Check setting of distribution valves to be sure steam goes only to those branch lines ready to receive it.
2. Open main supply valve slowly and in stages to control steam flow volume and provide gradual heatup of the line.
3. Watch discharge at drip valves. Do not close drip valves until warmup condensate has been discharged (except for next item).
4. Condensate should not be drained from drip pockets. An accumulation of condensate is necessary in drip pockets as they are in the line to do the following:

    A. To let condensate escape by gravity from the fast moving steam,

    B. To store condensate until the pressure differential is great enough for the steam trap to discharge,

   C. Provide storage of condensate until there is positive pressure
      in the line,

   D. Provide static head enabling trap to discharge before a posi-
      tive pressure exists in the line.

5. Check to see that line pressure comes up to the required operating pressure.

6. Check operation of all steam traps draining condensate from
   line to be sure they are operating properly (check temperature,
   discharge, etc.).

AUTOMATIC SPRINKLER SYSTEMS

Wet Sprinkler Systems

A *wet sprinkler system* (Fig. 178) is described by the National
Board of Fire Underwriters as "a system employing automatic sprinklers attached to a piping system containing water and connected to a
water supply so that water discharges immediately from sprinklers
opened by a fire." A vital component of a wet sprinkler system is the
alarm check valve, also called *wet sprinkler valve.* Its function is to
direct water to alarm devices and sound the sprinkler alarm. *It does
NOT control the flow of water into the system.*

The alarm check valve is located in the pipe riser at the point the
water line enters the building. An underground valve with an indicator post is usually located a safe distance outside the building. In
design and operation the alarm check valve is a globe type check valve.
A groove cut in the seat is connected by passage to a threaded outlet
on the side of the valve body. When the valve lifts the water can
flow to the outlet. There is also a large drain port above the seat which
is connected to the *drain valve.* Two additional ports, one above and
one below the seat, allow the attachment of pressure gauges.

A plug or stop-cock type valve called the *alarm control cock* is
connected to the outlet from the seat groove. This cock controls the
water flow to the alarm devices, allowing their silencing. This cock
MUST be in the alarm position when the system is "set," or the alarm
check valve will be unable to perform its function of sounding an alarm.

Operation Of Wet Sprinkler Alarm System

When a sprinkler head opens, or for any reason water escapes from
a wet sprinkler system, the flow through the alarm check valve causes

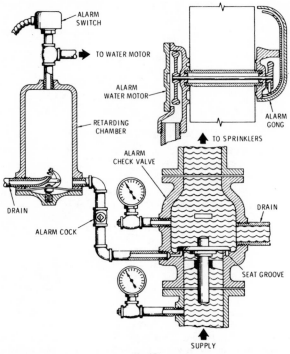

Fig. 178

the check valve disc to lift. Water entering the seat groove flows through the alarm cock to the retarding chamber. The function of the retarding chamber is to avoid unnecessary alarms which might be caused by slight leakage. It will allow a small volume water flow without actuating the alarm. When there is a large flow, as occurs when a sprinkler opens, the chamber is quickly filled and pressure closes the diaphragm-actuated drain valve in the bottom of the retarding chamber. The electrical alarm is then actuated by the pressure of the water. The water also flows to the water motor, causing it to be rotated and whirl hammers inside the alarm gong, thus mechanically sounding the bell alarm.

Placing A Wet Sprinkler System In Service

1. Check system to be sure it is ready to be filled with water. If the system has been shut down because a head has opened, be sure the head has been replaced with one of proper rating.
2. Open "vent valves"—located at high points.
3. Place "alarm cock" in CLOSED position. This will prevent sounding of alarm while flowing water fills system.
4. Place "drain valve" in nearly closed position. A trickle of water should flow from drain valve during filling.
5. Open "indicator post valve" slowly. When the system has filled there will be a quieting of the sound of rushing water. Open valve full open then back off ¼ turn.
6. Observe the water flow at the vent valves. When a steady flow of water occurs (no air), close vent valves.
7. Check the water flow by quickly opening the "drain valve" and closing it. The water pressure should drop about 10 pounds when the valve is opened and immediately return to full pressure when valve is closed. Excessive pressure drop indicates insufficient water flow.
8. Open "alarm cock"—system is now in SET condition.
9. Test—Open "drain valve" several turns. Water should flow sounding the mechanical gong alarm and the electrical alarm.
10. Close "drain valve"—If the alarms have functioned properly the system is operational.

Dry-Pipe Sprinkler Systems

A *dry-pipe sprinkler system* is described by the National Board of Fire Underwriters as: "a system employing automatic sprinklers attached to a piping system containing air under pressure, the release of which as from the opening of a sprinkler permits the water pressure to open a valve known as a *dry-pipe valve*. The water then flows into the piping system and out the open sprinkler."

The dry-pipe valve is located in the pipe riser at the point the water line enters the building. An underground valve with indicator post is usually located a safe distance outside the building.

The dry-pipe valve is a dual style valve, as both an air valve and a water valve are contained inside its body. These two internal valves

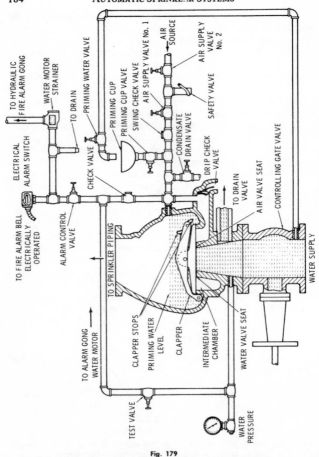

Fig. 179

may be separate units, one positioned above the other, or they may be combined in a single unit, one within the other. The function of the air valve is to retain the air in the piping system and to hold the water valve closed, thus restraining the flow of water. As long as sufficient air pressure is maintained in the system the air valve can do this since it has a larger surface area than the water valve. The air in the system acting on this large surface provides enough force to hold the water valve closed against the pressure of the water.

Operation of Dry-Pipe Sprinkler Systems

When a sprinkler head opens, or for any reason air escapes from a dry pipe system, the air pressure above the internal air valve is reduced. When the air pressure falls to the point its force is exceeded by the force of the water below the internal valve, both valves are thrown open. This allows an unobstructed flow of water through the "Dry Pipe Valve" into the system piping and to the open sprinklers. As the valves are thrown open water fills the "Intermediate Chamber." To avoid unnecessary alarms the chamber is equipped with a "Velocity Drip Valve." This drip valve is normally open to the atmosphere and allows drainage of any slight water leakage past the internal water valve seat.

When a sprinkler head opens and falling air pressure allows the internal valves to be thrown open, the intermediate chamber instantly fills with water and the "Velocity Drip Valve" is forced closed. The water now under pressure in the chamber flows to the electrical alarm switch and to the alarm gong.

Placing a Dry-Pipe Sprinkler System in Service

1. Close the valve controlling water flow to the system. This may be located in the riser or it may be an underground valve with an indicator post. If a fire has occurred and water is flowing from opened sprinklers approval of the person in authority must be obtained before closing the valve.

2. Open the "Drain Valve" and allow the water to drain from the sprinkler system piping.

3. Open all "Vent" and "Drain" valves throughout the system. Vents will be located at the high points and drains at all trapped and low portions of the piping system.

4. Manually push open the "Velocity Drip Valve." Also open the "Drain" valve for the "Dry Pipe Valve Body," if one is provided.

5. Remove the cover plate from the "Dry Pipe Valve" and carefully clean the rubber facings and seat surfaces of the internal

air and water valves. DO NOT use rags or abrasive wiping materials. Wipe the seats clean with the bare fingers.

6. Unlatch the "Clapper" and carefully close the internal air and water valves.

7. Replace the "Dry Pipe Valve" cover and close the "Drain" valve, if one is provided.

8. Open the "Priming Cup Valve" and "Priming Water Valve" to admit priming seal water into the "Dry Pipe Valve" to the level of the pipe connection. The priming water provides a more positive seal to prevent air from escaping past the air valve seat into the intermediate chamber.

9. Drain excess water by opening "Condensate Drain Valve." Close tightly when water no longer drains from valve.

10. Open "Air Supply Valve" and admit air to build up a few pounds of pressure in the system.

11. Check all open vents and drains throughout the system to be sure all water has been forced from the low points. As soon as dry air exhausts at the various open points the openings should be closed. Close "Air Supply Valve."

12. Replace any open sprinklers with new sprinkler heads of the proper rating.

13. Open "Air Supply Valve" and allow system air pressure to build up to the required pressure. The air pressure required to keep the internal valves closed varies directly with water supply pressure. Consult pressure setting tables.

14. Open the system water supply valve slightly to obtain a small flow of water to the "Dry Pipe Valve."

15. When water is flowing clear at the drain valve, slowly close it allowing water pressure to gradually build up below the internal water valve as observed on the water pressure gage.

16. When water pressure has reached the maximum below the internal water valve, open the supply valve to full open position. Back off valve about a ¼ turn from full open.

17. Test the alarms—Open the "Test Valve" or if system has a three position test cock place cock in TEST position. Water should flow to the electrical alarm switch and also to the alarm gong water motor.

18. If alarms have functioned properly close the "Test Valve" or place the three position test cock in ALARM position.

The alarm test is a test of the functioning of the alarm system only and does not indicate the condition of the "Dry Pipe Valve." The "Dry Pipe Valve" operation is tested by opening a vent valve to allow air in the piping system to escape, causing the "Dry Pipe Valve" to trip. It must then be reset, going through the procedure listed above.

Carpentry

Commercial Lumber Sizes

Two words *timber* and *lumber* are commonly used to describe the principal material used by carpenters. In the early stages of lumber production the word timber is usually applied to wood while in its natural state. Wood after cutting and sawing into standard commercial pieces is called *lumber*.

Stock lumber may be *green,* meaning the wood contains a large percentage of its natural moisture, or it may be *seasoned.* Seasoning is the process, either naturally or by exposure to heat, of removing about 85% of the moisture contained in freshly cut timber. Lumber is usually classified according to the three types into which it is rough sawed. These are: "Dimension Stock," which is 2 inches thick and from 4 to 12 inches wide. "Timbers," which are 4 to 8 inches thick and 6 to 10 inches wide. "Common Boards," which are 1 inch thick and 4 to 12 inches wide.

Rough lumber is *dressed* or *surfaced* by removing about ⅛ inch from each side and about ⅜ inch from the edges. This planing operation is called *dressing* or *surfacing,* and the letter "D" for dressed or "S" for surfaced are used to indicate how many sides or edges are planed. For example D.1.S. or S.1.S. indicates one side has been planed. Lumber planed on both sides and edges is designated S.4.S.

Building Layout

The first step in building layout is to establish by measurement from boundary markers or other reliable positions the wall line for one side of the building, and one building corner location point on the line. A stake is driven into the ground at this point and a nail driven into the top of the stake to accurately mark the corner point. The other corner point stakes are then located by measurement and by (squaring) the corners.

Table 30. New and Old Standard Lumber Sizes

Lumber Classification	Nominal Size		Actual S4S Size		Old S4S Size	
	Thickness	Width	Thickness	Width	Thickness	Width
Dimension	2″	4″	1½″	3½″	1⅝″	3⅝″
	2″	6″	1½″	5½″	1⅝″	5⅝″
	2″	8″	1½″	7¼″	1⅝″	7½″
	2″	10″	1½″	9¼″	1⅝″	9½″
	2″	12″	1½″	11¼″	1⅝″	11½″
Timbers	4″	6″	3½″	5½″	3⅝″	5⅝″
	4″	8″	3½″	7¼″	3⅝″	7½″
	4″	10″	3½″	9¼″	3⅝″	9½″
	6″	6″	5½″	5½″	5⅝″	5⅝″
	6″	8″	5½″	7¼″	5⅝″	7½″
	6″	10″	5½″	9¼″	5⅝″	9½″
	8″	8″	7¼″	7¼″	7½″	7½″
	10″	10″	9¼″	9¼″	9½″	9½″
Common Boards	1″	4″	¾″	3½″	$\frac{25}{32}$″	3⅝″
	1″	6″	¾″	5½″	$\frac{25}{32}$″	5⅝″
	1″	8″	¾″	7½″	$\frac{25}{32}$″	7⅝″
	1″	10″	¾″	9¼″	$\frac{25}{32}$″	9½″
	1″	12″	¾″	11¼″	$\frac{25}{32}$″	11½″

The corners are squared by use of the 3-4-5 triangle measurement system. Measurements are made along the sides in multiples of 3 and 4, and along the diagonal in multiples of 5. The 6-8-10 and 9-12-15 combination measurements are often used. To further assure square corners the diagonals are checked by measurement to see if they are the same length.

After the corner stakes are accurately located *batter board* stakes are driven in at each corner about 4 feet beyond the building lines, and the batter boards are attached. Batter boards are the usual method used to retain the outline of a building layout. The height of the boards may also be positioned to conform to the height of above grade foundation walls. A line is held across the top of opposite boards at the corners and adjusted using a plumb bob so that it is exactly over the nails in the stakes. Saw kerfs are cut where the lines touch the boards so that accurate line locations are assured after the corner stakes are removed.

Wood-Frame Building Construction

While the details of wood-frame building construction may vary in different localities, the fundamental principles are the same. The following illustrations show established methods of construction and accepted practices used in wood-frame building construction.

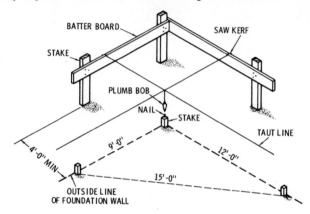

Fig. 180. Layout stakes and batter boards.

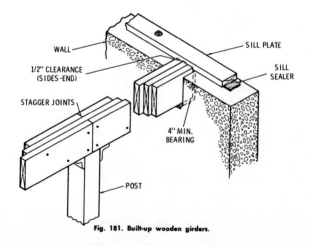

Fig. 181. Built-up wooden girders.

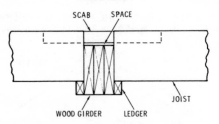

Fig. 182. Connecting scab used to tie joists together.

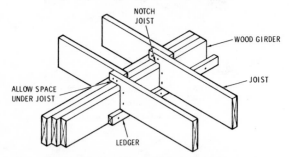

Fig. 183. Floor joists notched to fit over girder.

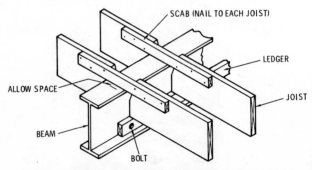

Fig. 184. Floor joists resting on wooden ledger fastened to "I" beam girder.

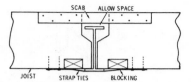

Fig. 185. Floor joists resting directly on "I" beam girder and connected at top with scab board

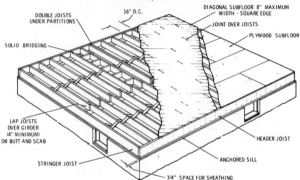

Fig. 186. Platform construction—details of floor joists and subflooring.

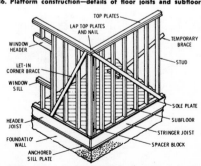

Fig. 187. Platform construction—wall stud details with let-in bracing and double top plates.

CARPENTRY

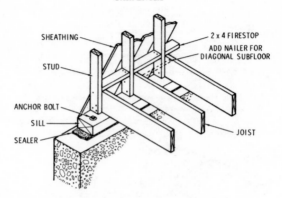

SHEATHING

2 x 4 FIRESTOP

ADD NAILER FOR
DIAGONAL SUBFLOOR

STUD

ANCHOR BOLT

SILL

SEALER

JOIST

Fig. 188. Balloon-frame construction—wall studs and floor joists rest on anchored sill.

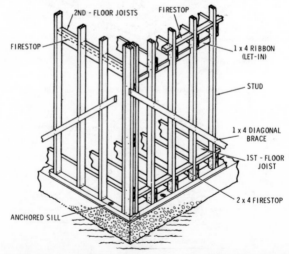

2ND - FLOOR JOISTS   FIRESTOP

FIRESTOP

1 x 4 RIBBON
(LET-IN)

STUD

1 x 4 DIAGONAL
BRACE

1ST - FLOOR
JOIST

2 x 4 FIRESTOP

ANCHORED SILL

Fig. 189. Balloon-frame construction—second-floor joists rest on 1" x 4" ribbons that have
been let into the wall studs. Fire stops prevent spread of fire through open wall passages.

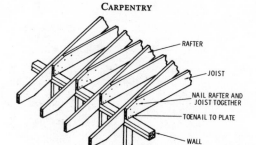

Fig. 190. Ceiling joist and rafter construction.

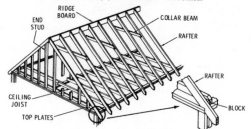

Fig. 191. Roof frame construction.

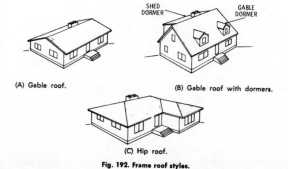

(A) Gable roof.

(B) Gable roof with dormers.

(C) Hip roof.

Fig. 192. Frame roof styles.

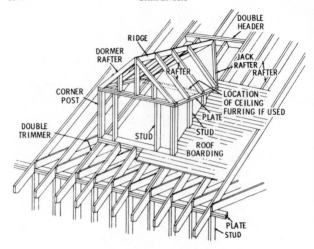

Fig. 193. Dormer frame construction.

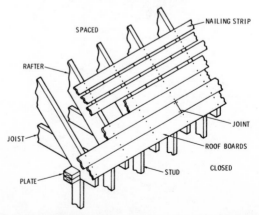

Fig. 194. Board roof sheathing—spaced for wood shingles or closed for asphalt shingles.

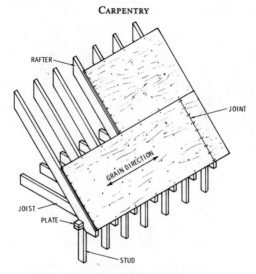

Fig. 195. Plywood roof sheathing.

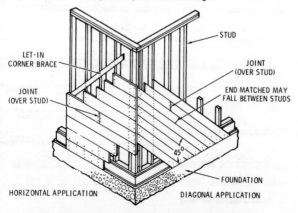

Fig. 196. Horizontal or diagonal applied board wall sheathing.

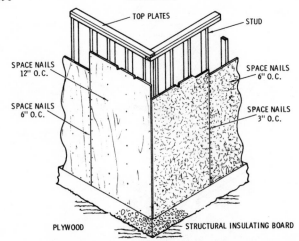

TOP PLATES

STUD

SPACE NAILS 12" O.C.

SPACE NAILS 6" O.C.

SPACE NAILS 6" O.C.

SPACE NAILS 3" O.C.

PLYWOOD

STRUCTURAL INSULATING BOARD

Fig. 197. Plywood or insulating board wall sheathing.

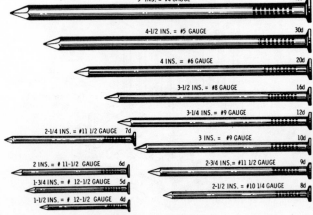

5 INS. = #4 GAUGE   40d

4-1/2 INS. = #5 GAUGE   30d

4 INS. = #6 GAUGE   20d

3-1/2 INS. = #8 GAUGE   16d

3-1/4 INS. = #9 GAUGE   12d

2-1/4 INS. = #11 1/2 GAUGE   7d

3 INS. = #9 GAUGE   10d

2 INS. = # 11-1/2 GAUGE   6d

2-3/4 INS. = #11 1/2 GAUGE   9d

1-3/4 INS. = # 12-1/2 GAUGE   5d

2-1/2 INS. = #10 1/4 GAUGE   8d

1-1/2 INS. = # 12-1/2 GAUGE   4d

Chart 5. Common Wire Nails

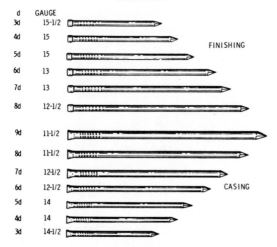

d	GAUGE		
3d	15-1/2		
4d	15		
5d	15		FINISHING
6d	13		
7d	13		
8d	12-1/2		
9d	11-1/2		
8d	11-1/2		
7d	12-1/2		
6d	12-1/2		CASING
5d	14		
4d	14		
3d	14-1/2		

Chart 6. The finishing nail is larger in diameter than a casing nail of equal length, in addition to having a different head shape.

Table 31. Standard Wood Screw Dimensions

Screw Gauge	Screw Diameter	Head Diameter		
		Flat	Round	Oval
0	0.060	0.112	0.106	0.112
1	0.073	0.138	0.130	0.138
2	0.086	0.164	0.154	0.164
3	0.099	0.190	0.178	0.190
4	0.112	0.216	0.202	0.216
5	0.125	0.242	0.228	0.242
6	0.138	0.268	0.250	0.268
7	0.151	0.294	0.274	0.294
8	0.164	0.320	0.298	0.320
9	0.177	0.346	0.322	0.346
10	0.190	0.371	0.346	0.371
11	0.203	0.398	0.370	0.398
12	0.216	0.424	0.395	0.424
13	0.229	0.450	0.414	0.450
14	0.242	0.476	0.443	0.476
15	0.255	0.502	0.467	0.502

FLAT HEAD ROUND HEAD OVAL HEAD

MEASURE OVERALL

MEASURE FROM BOTTOM OF THE HEAD

MEASURE FROM EDGE OF HEAD

Table 31. Standard Wood Screw Dimensions (Continued)

Screw Gauge	Screw Diameter	Head Diameter		
		Flat	Round	Oval
16	0.268	0.528	0.491	0.528
17	0.282	0.554	0.515	0.554
18	0.394	0.580	0.524	0.580
20	0.321	0.636	0.569	0.636
22	0.347	0.689	0.611	0.689
24	0.374	0.742	0.652	0.742
26	0.400	0.795	0.694	0.795
28	0.426	0.847	0.735	0.847
30	0.453	0.900	0.777	0.900

Table 32. Standard Machine Bolt Dimensions
(U.S. Standard or National Coarse Threads)

HEX HEAD   SQUARE HEAD

Diameter	Number of Threads per inch (National Coarse Thread)	Head	Head	Head
1/4	20	3/8	13/32	1/2
5/16	18	1/2	35/64	43/64
3/8	16	9/16	5/8	3/4
7/16	14	5/8	11/16	53/64
1/2	13	3/4	53/64	1
9/16	12	7/8	31/32	1 5/32
5/8	11	15/16	1 3/32	1 1/4
3/4	10	1 1/8	1 15/64	1 1/2
7/8	9	1 5/16	1 29/64	1 47/64
1	8	1 1/2	1 23/32	1 63/64
1 1/8	7	1 11/16	1 55/64	2 15/64
1 1/4	7	1 7/8	2 1/16	2 31/64
1 3/8	6	2 1/16	2 17/64	2 47/64
1 1/2	6	2 1/4	2 3/64	2 63/64
1 5/8	5 1/2	2 7/16	2 11/16	3 15/64
1 3/4	5	2 5/8	2 57/64	3 31/64
1 7/8	5	2 13/16	3 3/32	3 47/64
2	4 1/2	3	3 5/16	3 63/64

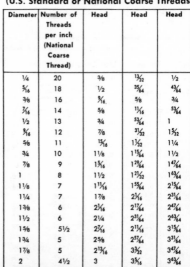

Table 33. Schedule of Common Wire Nail Use in Wood Frame Building Construction

Joining	Nailing method	Number	Size	Nails Placement
Header to joist	End-nail	3	16d	
Joist to sill or girder	Toenail	2	10d or	
		3	8d	
Header and stringer joist to sill	Toenail		10d	16 in. on center
Bridging to joist	Toenail each end	2	8d	At each joist
Ledger strip to beam, 2 in. thick		3	16d	
Subfloor, boards:				
1 by 6 in. and smaller		2	8d	To each joist
1 by 8 in.		3	8d	To each joist
Subfloor, plywood:				
At edges			8d	6 in. on center
At intermediate joists			8d	8 in. on center
Subfloor (2 by 6 in., T&G) to joist or girder	Blind-nail (casing) and face-nail	2	16d	
Soleplate to stud, horizontal assembly	End-nail	2	16d	At each stud
Top plate to stud	End-nail	2	16d	
Stud to soleplate	Toenail	4	8d	
Soleplate to joist or blocking	Face-nail		16d	16 in. on center
Doubled studs	Face-nail, stagger		10d	16 in. on center
End stud of intersecting wall to exterior wall stud	Face-nail		16d	16 in. on center
Upper top plate to lower top plate	Face-nail		16d	16 in. on center
Upper top plate, laps and intersections	Face-nail	2	16d	
Continuous header, two pieces, each edge			12d	12 in. on center
Ceiling joist to top wall plates	Toenail	3	8d	

Table 33. Schedule of Common Wire Nail Use in Wood Frame Building Construction (Continued)

Item	Nailing method	Number	Size	Placement
Ceiling joist laps at partition	Face-nail	4	16d	
Rafter to top plate	Toenail	2	8d	
Rafter to ceiling joist	Face-nail	5	10d	
Rafter to valley or hip rafter	Toenail	3	10d	
Ridge board to rafter	End-nail	3	10d	
Rafter to rafter	Toenail	4	8d	
Rafter to rafter through ridge board	Edge-nail	1	10d	
Collar beam to rafter:				
2 in. member	Face-nail	2	12d	
1 in. member	Face-nail	3	8d	
1-in. diagonal let-in brace to each stud and plate (4 nails at top)	Face-nail	2	8d	
Built-up corner studs:				
Studs to blocking	Face-nail	2	10d	Each side
Intersecting stud to corner studs	Face-nail		16d	12 in. on center
Built-up girders and beams, three or more members	Face-nail		20d	32 in. on center, each side
Wall sheathing:				
1 by 8 in. or less, horizontal	Face-nail	2	8d	At each stud
1 by 6 in. or greater, diagonal	Face-nail	3	8d	At each stud
Wall sheathing, vertically applied plywood:				
3/8 in. and less thick	Face-nail		6d	6 in. edge
1/2 in. and over thick	Face-nail		8d	12 in. intermediate
Wall sheathing, vertically applied fiberboard:				
1/2 in. thick	Face-nail			1½ in. roofing nail } 3 in. edge and
25/32 in. thick	Face-nail			1¾ in. roofing nail } 6 in. intermediate
Roof sheathing, boards, 4-, 6-, 8-in. width	Face-nail	2	8d	At each rafter
Roof sheathing, plywood:				
3/8 in. and less thick	Face-nail		6d	6 in. edge and 12 in. intermediate
1/2 in. and over thick	Face-nail		8d	6 in. edge and 12 in. intermediate

Electricity

Because of the potential danger, ever present with electrical energy, a basic requirement when working with electricity is that there be no guess work or chance taking. Therefore, activities in the area of electrical work should be restricted to those things with which one has experience or about which one has specific knowledge or understanding.

Electrical Terms

Electromotive Force—The force which causes electricity to flow when there is a difference of potential between two points. The unit of measurement is the volt.

Direct Current—The flow of electricity in one direction. This is commonly associated with continuous direct current which is nonpulsating, as from a storage battery.

Alternating Current—The flow of electricity that is continuously reversing or alternating in direction resulting in a regularly pulsating flow.

Voltage—The value of the electromotive force in an electrical system. It may be compared to pressure in a hydraulic system.

Amperage—The quantity and rate of flow in an electrical system. It may be compared to the volume of flow in a hydraulic system.

Resistance—The resistance offered by materials to the movement of electrons, commonly referred to as the flow of electricity. The unit of measurement commonly used is the *ohm*.

Cycle—The interval or period during which alternating current, (using zero as a starting point) increases to maximum force in a positive direction, reverses and decreases to zero, then increases to maximum force in a negative direction, then reverses again and decreases to zero value. One cycle of such a flow is visually represented in Fig. 198.

Frequency—The number of complete cycles per second of the alternating current flow. The most widely used alternating current fre-

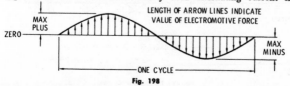

Fig. 198

quency is 60 cycles per second. This is the number of complete cycles per second, thus the pulsation rate is twice this or 120 pulses per second. Frequency is now specified as so many hertz. The term (hertz) is defined as cycles per second and is abbreviated Hz.

Phase—The word (phase) applies to the number of current surges that flow simultaneously in an electrical circuit. Fig. 198 is a graphic representation of single-phase alternating current. The single line represents a current flow that is continuously increasing or decreasing in value.

Three-phase current has three separate surges of current flowing together. In any given instant however, their values differ as the peaks and valleys of the pulsations are spaced equally apart (Fig. 199). The waveforms are lettered A, B, and C to represent the alternating current flow for each phase during a complete cycle. In three phase current flow, any one current pulse is always one third of a cycle out of matching with another.

Watt—The *watt* is the electrical unit of power, or the rate of doing work. One watt represents the power that is used when one ampere of current flows in an electrical circuit with a voltage or pressure of one volt.

Watt Hour—The *watt hour* expresses watts in time measurement of hours. For example, if a 100-watt lamp is in operation for a 2-hour period, it will consume 200 watt hours of electrical energy.

Kilowatt Hour—One *kilowatt* is equal to 1000 watts. One kilowatt hour is the electrical energy expended at the rate of one kilowatt (1000 watts) over a period of one hour.

Electrical Calculations

Most simple electrical calculations associated with common electrical power circuits involve the use of two basic formulas. These are the

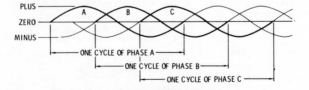

Fig. 199 Three-phase Current

Ohms law formula and the basic *electrical power* formula. By substitution of known values into these formulas, and their rearrangements, unknown values may be easily determined.

Ohm's Law

This is the universally used electrical law stating the relationship of current, voltage, and resistance. This is done mathematically by the formula shown below. Current is stated in *amperes* and abbreviated I. Resistance is stated in *ohms* and abbreviated R, and voltage in *volts* abbreviated E.

$$Current = \frac{Voltage}{Resistance} \quad or \quad I = \frac{E}{R}$$

The rearrangement of values gives two other forms of the same equation as follows:

$$R = \frac{E}{I} \quad and \quad E = I \times R$$

Example

An ammeter placed in a 110-volt circuit indicates a current flow of 5 amperes, what is the resistance of the circuit?

$$R = \frac{E}{I} \quad or \quad R = \frac{110}{5} \quad or \quad R = 22 \ ohms$$

Power Formula

This formula indicates the rate at any given instant at which work is being done by current moving through a circuit. Voltage and Amperes are abbreviated E and I as in Ohm's law, and watts are abbreviated W.

$$Watts = Voltage \times Amperes \quad or \quad W = E \times I$$

The two other forms of the formula obtained by rearrangement of the values are:

$$E = \frac{W}{I} \quad and \quad I = \frac{W}{E}$$

Example

Using the same values as used in the example above, 5 amperes flowing in a 110 volt circuit, how much power is consumed?

$$W = E \times I \quad or \quad W = 110 \times 5 \quad or \quad W = 550 \ watts$$

Example

A 110-volt appliance is rated at 2000 watts, can this appliance be plugged into a circuit fused at 15 amps?

$$I = \frac{W}{E} \quad or \quad I = \frac{2000}{110} \quad or \quad I = 18.18 \ amperes$$

Obviously the fuse would blow if this appliance were plugged into the circuit.

Electrical Circuits

An electrical circuit is composed of elements such as lamps, switches, motors, resistors, wires, cables, batteries or other voltage sources, etc. All are conductors or conducting devices which form an electrical path called a circuit. To represent the various elements of a circuit on paper, circuit diagrams composed of lines and symbols are used. These are called *schematic diagrams,* and symbols used to represent the circuit elements, including the voltage source, are standardized. Some of the symbols commonly used in industrial applications are shown in Table 34.

Table 34. Symbols Used in Industrial Applications

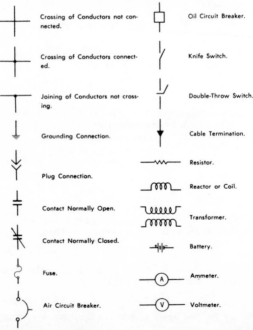

Symbol	Description	Symbol	Description
	Crossing of Conductors not connected.		Oil Circuit Breaker.
	Crossing of Conductors connected.		Knife Switch.
	Joining of Conductors not crossing.		Double-Throw Switch.
	Grounding Connection.		Cable Termination.
	Plug Connection.		Resistor.
	Contact Normally Open.		Reactor or Coil.
	Contact Normally Closed.		Transformer.
	Fuse.		Battery.
	Air Circuit Breaker.		Ammeter.
			Voltmeter.

Electrical circuits may be classified as *series* circuits, *parallel* circuits, or a combination of series and parallel circuits. A series circuit is one where all parts of the circuit are electrically connected end to end. The current flows from one terminal of the power source through each element and to the other power-source terminal. The same amount of current flows in each part of the circuit. An example of a series circuit is illustrated in Fig. 200.

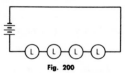

Fig. 200

A parallel circuit is one where each element is so connected that it has direct flow to both terminals of the power source. The voltage across any element in a parallel circuit is equal to the voltage of the source, or power supply. An example of a parallel circuit is illustrated in Fig. 201.

The relationship of values in series and parallel circuits using Ohm's law and the power formula are illustrated and compared in the following examples.

Series Circuit	*Parallel Circuit*

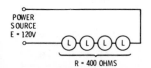

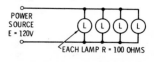

Current Flow Thru Circuit

$$I = \frac{E}{R} = \frac{120}{400} = .3 \ amp$$

Voltage Across One Lamp

$E = IR = .3 \times 100 = 30 \ volts$

Current Flow Thru One Lamp

$$I = \frac{E}{R} = \frac{30}{100} = .3 \ amp$$

Power Used By One Lamp

$W = EI = 30 \times .3 = 9 \ watts$

Current Flow Thru Circuit

$$I = \frac{E}{R} = \frac{120}{25} = 4.8 \ amps$$

Current Flow Thru One Lamp

$$I = \frac{E}{R} = \frac{120}{100} = 1.2 \ amps$$

Voltage Across One Lamp

$E = IR = 1.2 \times 100 = 120 \ volts$

Power Used By One Lamp

$W = EI = 120 \times 1.2 = 144 watts$

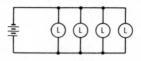

Fig. 201

Power Used By Circuit	Power Used By Circuit
$W = EI = 120 \times .3 = 36 \ watts$	$W = EI = 120 \times 4.8 = 576 \ watts$

Electrical Wiring

The term *electrical wiring* is applied to the installation and assembly of electrical conductors. The size of the wire used for electrical conductors is specified by gauge number according to the American Wire Gauge System. The usual manner of designation is by abbreviation AWG. Table 35 lists the AWG numbers and the corresponding specifications using the *mil* unit to designate a 0.001-inch measurement.

The *circular mil* unit used in the table is a measurement of cross-sectional area based on a circle one mil in diameter.

Table 35. Size, Resistance, and Weight of Standard Annealed Copper Wire

Size of Wire, AWG	Diameter of Wire Mils	Cross Section Circular Mils	Resistance, Ohms per 1000 Ft. at 68 F or 20 C	Weight, Pounds per 1000 Ft.
0000	460	212,000	0.0500	641
000	410	168,000	.062	508
00	365	133,000	.078	403
0	325	106,000	.098	319
1	289	83,700	.124	253
2	258	66,400	.156	201
3	229	52,600	.197	159
4	204	41,700	.248	126
5	182	33,100	.313	100
6	162	26,300	.395	79.5
7	144	20,800	.498	63.0
8	128	16,500	.628	50.0
9	144	13,100	.792	39.6

Table 35. Size, Resistance, and Weight of Standard Annealed Copper Wire (Continued)

Size of Wire, AWG	Diameter of Wire Mils	Cross Section Circular Mils	Resistance, Ohms per 1000 Ft. at 68 F or 20 C	Weight, Pounds per 1000 Ft.
10	102	10,400	0.998	31.4
11	91	8,230	1.26	24.9
12	81	6,530	1.59	19.8
13	72	5,180	2.00	15.7
14	64	4,110	2.53	12.4
15	57	3,260	3.18	9.86
16	51	2,580	4.02	7.82
17	45	2,050	5.06	6.20
18	40	1,620	6.39	4.92
19	36	1,290	8.05	3.90
20	32	1,020	10.15	3.09
21	28.5	810	12.80	2.45
22	25.3	642	16.14	1.94
23	22.6	509	20.36	1.54
24	20.1	404	25.67	1.22
25	17.9	320	32.37	0.970
26	15.9	254	40.81	.769
27	14.2	202	51.47	.610
28	12.6	160	64.90	.484
29	11.3	127	81.83	.384
30	10.0	101	103.2	.304
31	8.9	79.7	130.1	.241
32	8.0	63.2	164.1	0.191

Table 36. Current Capacities of Copper Wires (Amperes)

Wire Size	In Conduit or Cable		In Free Air		Weatherproof Wire
	Type RHW*	Type TW, R*	Type RHW*	Type TW, R*	
14	15	15	20	20	30
12	20	20	25	25	40
10	30	30	40	40	55
8	45	40	65	55	70
6	65	55	95	80	100
4	85	70	125	105	130
3	100	80	145	120	150
2	115	95	170	140	175

Table 36. Current Capacities of Copper Wires (Amperes) (Continued)

Wire Size	In Conduit or Cable		In Free Air		Weather-proof Wire
	Type RHW*	Type TW. R*	Type RHW*	Type TW, R*	
1	130	110	195	165	205
0	150	125	230	195	235
00	175	145	265	225	275
000	200	165	310	260	320

* Types "RHW", "TW", or "R" are identified by markings on outer cover

Table 37. Adequate Wire Sizes—Weatherproof Copper Wire

Load in Building Amperes	Distance in Feet from Pole to Building	*Recommended Size of Feeder Wire for Job
Up to 25 amperes, 120 volts	Up to 50 feet	No. 10
	50 to 80 feet	No. 8
	80 to 125 feet	No. 6
20 to 30 amperes, 240 volts	Up to 80 feet	No. 10
	80 to 125 feet	No. 8
	125 to 200 feet	No. 6
	200 to 350 feet	No. 4
30 to 50 amperes, 240 volts	Up to 80 feet	No. 8
	80 to 125 feet	No. 6
	125 to 200 feet	No. 4
	200 to 300 feet	No. 2
	300 to 400 feet	No. 1

* These sizes are recommended to reduce "voltage drop" to a minimum

Switches are the most widely used of all electric wiring devices. They are connected in series with the devices they control and allow current to flow when closed and interrupt current flow when open. One of the most common of switch applications is the control of one or more lamps from a single location. The schematic diagram for such a circuit is illustrated in Fig. 202 and a sketch of actual wiring connections is shown in Fig. 203.

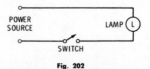

Fig. 202

Table 38. Circuit Wire Sizes for Individual Single-phase Motors

Horsepower of Motor	Volts	Approximate Starting Current Amperes	Approximate Full Load Current Amperes	Feet	LENGTH OF RUN IN FEET from Main Switch to Motor							
					25	50	75	100	150	200	300	400
1/4	120	20	5	Wire Size	14	14	14	12	10	10	8	6
1/3	120	20	5.5	Wire Size	14	14	14	12	10	8	6	6
1/2	120	22	7	Wire Size	14	14	12	12	10	8	6	6
3/4	120	28	9.5	Wire Size	14	12	12	10	8	6	4	4
1/4	240	10	2.5	Wire Size	14	14	14	14	14	14	12	12
1/3	240	10	3	Wire Size	14	14	14	14	14	14	12	10
1/2	240	11	3.5	Wire Size	14	14	14	14	14	12	12	10
3/4	240	14	4.7	Wire Size	14	14	14	14	14	12	10	10
1	240	16	5.5	Wire Size	14	14	14	14	14	12	10	10
1 1/2	240	22	7.6	Wire Size	14	14	14	14	12	10	8	8
2	240	30	10	Wire Size	14	14	14	12	10	10	8	6
3	240	42	14	Wire Size	14	12	12	12	10	8	6	6
5	240	69	23	Wire Size	10	10	10	8	8	6	4	4
7 1/2	240	100	34	Wire Size	8	8	8	8	6	4	2	2
10	240	130	43	Wire Size	6	6	6	6	4	4	2	1

Table 39. Types and usage of extension cords

	Type	Wire Size	Use
Ordinary Lamp Cord	POSJ SPT	No. 16 or 18	In residences for lamps or small appliances.
Heavy-duty—with thicker covering	S, SJ or SJT	No. 10, 12, 14 or 16	In shops, and outdoors for larger motors, lawn mowers, outdoor lighting, etc.

Table 40. Ability of cord to carry current (2 or 3-wire cord)

Wire Size	Type	Normal Load	Capacity Load
No. 18	S, SJ, SJT or POSJ	5.0 Amp. (600W)	7 Amp. (840W)
No. 16	S, SJ, SJT or POSJ	8.3 Amp. (1000W)	10 Amp. (1200W)
No. 14	S	12.5 Amp. (1500W)	15 Amp. (1800W)
No. 12	S	16.6 Amp. (1900W)	20 Amp. (2400W)

Table 41. Selecting the length of cord set

Light Load (to 7 amps.)	Medium Load (7-10 amps.)	Heavy Load (10-15 amps.)
To 25 Ft.—Use No. 18	To 25 Ft.—Use No. 16	To 25 Ft.—Use No. 14
To 50 Ft.—Use No. 16	To 50 Ft.—Use No. 14	To 50 Ft.—Use No. 12
To 100 Ft.—Use No. 14	To 100 Ft.—Use No. 12	To 100 Ft.—Use No. 10

NOTE: As a safety precaution be sure to use only cords which are listed by Underwriters' Laboratories. Look for the Underwriters' seal when you buy.

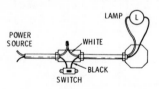

Fig. 203

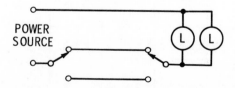

Fig. 204

To control lamps from two points requires a type switch called a three-way switch. It has three terminals, one of which is so arranged that current is carried through it to either of the other two. Its function is to connect one wire to either of two other wires. The diagram in Figure 204 shows a lamp circuit controlled from two points using three-way switches. The actual connection boxes and the wires in the cable between the boxes for the three-way switch circuit are shown in Fig. 205.

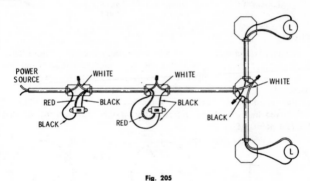

Fig. 205

Shop Geometry

Lines

When lines touch at one point they are said to be "tangent" to one another. Two lines that cross are said to "intersect." Two lines that are always the same distance apart are said to be "parallel."

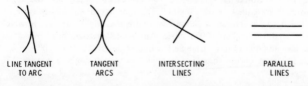

| LINE TANGENT TO ARC | TANGENT ARCS | INTERSECTING LINES | PARALLEL LINES |

Angles

There are four (4) common types of angles: straight, right, acute, and obtuse.

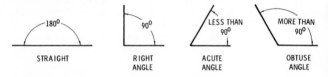

STRAIGHT   RIGHT ANGLE   ACUTE ANGLE   OBTUSE ANGLE

Circles

A circle is a closed curve on which all points are the same distance from a point (center) inside the circle. A line extending from the center to the closed curve line is the "radius." The straight line that passes through the center of the circle and ends at the opposite side of the closed curved line is the "diameter." An "arc" is a portion of the curved line. A "chord" refers to the straight line segment between the end points of an arc.

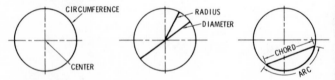

The distance around the circle is called the "circumference." The circumference is measured in standard linear units or in angular measure. The total angular distance around the circumference of a circle is 360 degrees.

Triangles

A triangle consists of three straight lines joined at the end points to form a closed flat shape. The lines are called "sides" and the angles formed are called "inside angles." Triangles are described by their included angles. The most commonly used is the "right" triangle in which one angles is a right (90°) angle. An "acute" triangle has three acute angles and an "obtuse" triangle has one angle greater than 90°.

The sum of the three inside angles of any triangle is always 180 degrees.

RIGHT
TRIANGLE

ACUTE
TRIANGLE

OBTUSE
TRIANGLE

Quadrilaterals

Four sided figures are called quadrilaterals. When all four angles are right angles, and two pair of sides are of equal length, the shape is a "rectangle." When all four angles are right angles, and the four sides are equal, it is called a square.

ALL ANGLES
90 DEGREES   OPPOSITE
SIDES EQUAL

RECTANGLE

ALL ANGLES
90 DEGREES   ALL SIDES
EQUAL

SQUARE

Regular Polygons

All plane figures with three or more sides are called "polygons." Polygons having equal sides and equal angles are called "regular polygons." The names given regular polygons imply the number of equal sides.

Number of Sides	Name	Number of Sides	Name
3	Triangle	7	Heptagon
4	Square	8	Octagon
5	Pentagon	9	Nonagon
6	Hexagon	10	Decagon

Geometrical Construction

Practical application of geometric principles are in the construction of parallel, perpendicular, and tangent lines, dividing of straight and curved lines, and the bisecting of angles. These principles have further broad application in the layout of geometric shapes.

Dividing A Line

A line may be divided into equal parts by first drawing a line at any angle and length to the given line from one end point. Second, stepping off on the angular line the same number of equal spaces as the line is to be divided into. Third, connecting the last point with the other end point on the given line. Fourth, drawing lines parallel to this connecting line. Dividing a line into three equal parts is illustrated below:

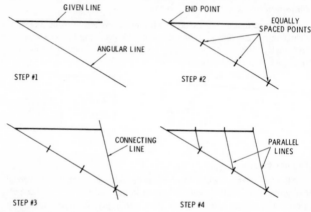

Erecting A Perpendicular Line

To erect a perpendicular or right angle line at a given point on a line, use the following three steps. First swing equal arcs from the given point to intersect the given line at points (a) and (b). Second, increase the radius by about one half and swing arcs from points (a) and (b) to locate points (c) and (d). Third, draw a line through point (c) and (d) and the given point.

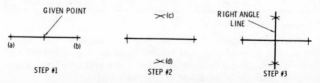

Polygon Construction

Square Inside A Circle
1. Draw a diameter line across circle.
2. Construct a perpendicular through center.
3. Connect the points of the diameter lines.

Square Outside A Circle
1. Draw a diameter line across circle.
2. Construct a perpendicular.
3. Construct tangent lines at the points of the diameter lines.

Hexagon Inside A Circle
1. Using the radius of the circle step off arcs around the circle.
2. Connect the points where the arcs intersect the circle.

Hexagon Outside A Circle
1. Using the radius of the circle step off arcs around the circle.
2. Construct a tangent line at each intersecting point.

Steps similar to these may be followed for constructing other polygons inside and outside circles.

Dimensions of Polygons and Circles

Triangle

$$E = \text{side} \times .57735$$
$$D = \text{side} \times 1.1547 = 2E$$
$$\text{Side} = D \times .866$$
$$C = E \times .5 = D \times .25$$

Square

$$E = \text{side} = D \times .7071$$
$$D = \text{side} \times 1.4142 = \text{Diagonal}$$
$$\text{Side} = D \times .7071$$
$$C = D \times .14645$$

Pentagon

$$E = \text{side} \times 1.3764 = D \times .809$$
$$D = \text{side} \times 1.7013 = E \times 1.2361$$
$$\text{Side} = D \times .5878$$
$$C = D \times .0955$$

Hexagon

$$E = \text{side} \times 1.7321 = D \times .866$$
$$D = \text{side} \times 2 = E \times 1.1547$$
$$\text{Side} = D \times .5$$
$$C = D \times .067$$

Octagon

$$E = \text{side} \times 2.4142 = D \times .9239$$
$$D = \text{side} \times 2.6131 = E \times 1.0824$$
$$\text{Side} = D \times .3827$$
$$C = D \times .038$$

Shop Trigonometry

Right-Angle Triangles

All triangles are made up of six parts—three sides and three angles. A right-angle triangle is one having one angle of 90 degrees. A 90-degree angle is termed a *right angle*. The three sides of a right-angle triangle are called the *side opposite, side adjacent* and *hypotenuse*. The hypotenuse is always the side directly across from, or opposite, the 90-degree angle. The other two sides are opposite one angle and adjacent to the other. Therefore they may be called either opposite or adjacent sides in reference to the two angles as illustrated below:

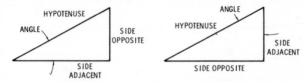

Trigonometric calculations employ numerical values called *trigonometric functions*. These values represent the ratios between the sides of triangles and are identified by the names *sine, cosine, tangent, cotangent, secant,* and *cosecant.* Each angle has a specific numerical value for each of its functions. These values are given in tables of trigonometric functions.

When one of the angles of a right-angle triangle (other than the 90° angle) and the length of one of the sides are known, the length of the other sides may be determind by use of the appropriate formula below:

$$\text{Length of side opposite} = \begin{cases} \text{hypotenuse} \times \text{sine} \\ \text{hypotenuse} \div \text{cosecant} \\ \text{side adjacent} \times \text{tangent} \\ \text{side adjacent} \div \text{cotangent} \end{cases}$$

$$\text{Length of side adjacent} = \begin{cases} \text{hypotenuse} \times \text{cosine} \\ \text{hypotenuse} \div \text{secant} \\ \text{side opposite} \times \text{cotangent} \\ \text{side opposite} \div \text{tangent} \end{cases}$$

$$\text{Length of hypotenuse} = \begin{cases} \text{side opposite} \times \text{cosecant} \\ \text{side opposite} - \text{sine} \\ \text{side adjacent} \times \text{secant} \\ \text{side adjacent} \div \text{cosine} \end{cases}$$

Right-Angle Triangles

When the length of two sides of a right angle triangle are known, the angles may be determined in two steps using the trigonometric functions of the angles.

Step #1

Using the appropriate formula calculate the numerical value of the function of the angle.

Step #2

Using a table of trigonometric functions, find the angle which corresponds to the function calculated by formula.

$$\text{Sine} = \frac{\text{Side Opposite}}{\text{Hypotenuse}} \qquad \text{Cotangent} = \frac{\text{Side Adjament}}{\text{Side Opposite}}$$

$$\text{Cosine} = \frac{\text{Side Adjacent}}{\text{Hypotenuse}} \qquad \text{Secant} = \frac{\text{Hypotenuse}}{\text{Side Adjacent}}$$

$$\text{Tangent} = \frac{\text{Side Opposite}}{\text{Side Adjacent}} \qquad \text{Cosecant} = \frac{\text{Hypotenuse}}{\text{Side Opposite}}$$

Example:

Use formula:
$$\text{Cosine} = \frac{\text{side adjacent}}{\text{hypotenuse}}$$

$$\text{Cosine} = \frac{6\frac{5}{8}}{7\frac{3}{4}} \text{ or } \frac{6.625}{7.750} = .85483$$

From trigonometric functions table: *Angle = 31° 15½″*

Sum of the Squares

When the length of two sides of a right angle triangle are known, the length of the third side may be determined by the use of the sum of the squares formula. It states that the square of the hypotenuse is equal to the sum of the squares of the other two sides. The basic formula is commonly written:

$$c^2 = a^2 + b^2$$

To find the length of the third side of a right-angle triangle when the lengths of two sides are known, the known values are substituted in the appropriate equation.

$$c = \sqrt{a^2 + b^2} \qquad a = \sqrt{c^2 - b^2} \qquad b = \sqrt{c^2 - a^2}$$

Example:

Find the length of c when a is 9 and b is 11. Use formula:

$$c = \sqrt{a^2 + b^2}$$
$$c = \sqrt{(9 \times 9) + (11 \times 11)} \text{ or } \sqrt{81 + 121} \text{ or } \sqrt{202}$$
$$c = \sqrt{202} \text{ or } 14.213 \text{ or approximately } 14\frac{7}{32}.$$

Note

Any triangle having sides with a 3-4-5 length ratio is a right-angle triangle.

Proof	Examples
$c^2 = a^2 + b^2$	6 — 8 — 10
$5^2 = 3^2 + 4^2$	12 — 16 — 20
$25 = 9 + 16$	15 — 20 — 25
$25 = 25$	18 — 24 — 30
	24 — 32 — 40
	30 — 40 — 50
	6- 8-10
	12-16-20
	15-20-25
	18-24-30
	24-32-40
	30-40-50

Natural Trigonometric Functions

Degree	Sine	Cosine	Tangent	Secant	Degree	Sine	Cosine	Tangent	Secant
0	.00000	1.0000	.00000	1.0999	46	.7193	.6947	1.0355	1.4395
1	.01745	.9998	.01745	1.0001	47	.7314	.6820	1.0724	1.4663
2	.03490	.9994	.03492	1.0006	48	.7431	.6691	1.1106	1.4945
3	.05234	.9986	.05241	1.0014	49	.7547	.6561	1.1504	1.5242
4	.06976	.9976	.06993	1.0024	50	.7660	.6428	1.1918	1.5557
5	.08716	.9962	.08749	1.0038	51	.7771	.6293	1.2349	1.5890
6	.10453	.9945	.10510	1.0055	52	.7880	.6157	1.2799	1.6243
7	.12187	.9925	.12278	1.0075	53	.7986	.6018	1.3270	1.6616
8	.1392	.9903	.1405	1.0098	54	.8090	.5878	1.3764	1.7013
9	.1564	.9877	.1584	1.0125	55	.8192	.5736	1.4281	1.7434
10	.1736	.9848	.1763	1.0154	56	.8290	.5592	1.4826	1.7883
11	.1908	.9816	.1944	1.0187	57	.8387	.5446	1.5399	1.8361
12	.2079	.9781	.2126	1.0223	58	.8480	.5299	1.6003	1.8871
13	.2250	.9744	.2309	1.0263	59	.8572	.5150	1.6643	1.9416
14	.2419	.9703	.2493	1.0306	60	.8660	.5000	1.7321	2.0000
15	.2588	.9659	.2679	1.0353	61	.8746	.4848	1.8040	2.0627
16	.2756	.9613	.2867	1.0403	62	.8829	.4695	1.8807	2.1300
17	.2924	.9563	.3057	1.0457	63	.8910	.4540	1.9626	2.2027
18	.3090	.9511	.3249	1.0515	64	.8988	.4384	2.0503	2.2812
19	.3256	.9455	.3443	1.0576	65	.9063	.4226	2.1445	2.3662
20	.3420	.9397	.3640	1.0642	66	.9135	.4067	2.2460	2.4586
21	.3584	.9336	.3839	1.0711	67	.9205	.3907	2.3559	2.5598
22	.3746	.9272	.4040	1.0785	68	.9272	.3746	2.4751	2.6695
23	.3907	.9205	.4245	1.0864	69	.9336	.3584	2.6051	2.7904
24	.4067	.9135	.4452	1.0946	70	.9397	.3420	2.7475	2.9238
25	.4226	.9063	.4663	1.1034	71	.9455	.3256	2.9042	3.0715
26	.4384	.8988	.4877	1.1126	72	.9511	.3090	3.0777	3.2361
27	.4540	.8910	.5095	1.1223	73	.9563	.2924	3.2709	3.4203
28	.4695	.8829	.5317	1.1326	74	.9613	.2756	3.4874	3.6279
29	.4848	.8746	.5543	1.1433	75	.9659	.2588	3.7321	3.8637
30	.5000	.8660	.5774	1.1547	76	.9703	.2419	4.0108	4.1336
31	.5150	.8572	.6009	1.1663	77	.9744	.2250	4.3315	4.4454
32	.5299	.8480	.6249	1.1792	78	.9781	.2079	4.7046	4.8097
33	.5446	.8387	.6494	1.1924	79	.9816	.1908	5.1446	5.2408
34	.5592	.8290	.6745	1.2062	80	.9848	.1736	5.6713	5.7588
35	.5736	.8192	.7002	1.2208	81	.9877	.1564	6.3138	6.3924
36	.5878	.8090	.7265	1.2361	82	.9903	.1392	7.1154	7.1853
37	.6018	.7986	.7536	1.2521	83	.9925	.12187	8.1443	8.2055
38	.6157	.7880	.7813	1.2690	84	.9945	.10453	9.5144	9.5668
39	.6293	.7771	.8098	1.2867	85	.9962	.08716	11.4301	11.474
40	.6428	.7660	.8391	1.3054	86	.9976	.06976	14.3007	14.335
41	.6561	.7547	.8693	1.3250	87	.9986	.05234	19.0811	19.107
42	.6691	.7431	.9004	1.3456	88	.9994	.03490	28.6363	28.654
43	.6820	.7314	.9325	1.3673	89	.9998	.01745	57.2900	57.299
44	.6947	.7193	.9657	1.3902	90	1.0000	Inf.	Inf.	Inf.
45	.7071	.7071	1.0000	1.4142	—	—	—	—	—

GEOMETRIC FORMULAS

Triangle

area $(A) = \dfrac{bh}{2}$

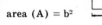

Sphere

area $(A) = 4\pi R^2$

$= \pi D^2$

volume $(V) = \dfrac{4}{3}\pi R^3$

$= 1/6\pi D^3$

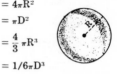

Square

area $(A) = b^2$

Rectangle

area $(A) = ab$

Cube

area $(A) = 6b^2$

volume $(V) = b^3$

Cone

area $(A) = \pi RS$

$= \pi R\sqrt{R^2 + h^2}$

volume $(V) = \dfrac{\pi R^2 h}{3}$

$= 1.047R^2 h$

$= 0.2618D^2 h$

Rectangular Solid

area $(A) =$
$2\,(ab + bc + ac)$

volume $(V) = abc$

Ring of Rectangular Cross Section

volume $(V) = \dfrac{\pi c}{4}(D^2 - d^2)$

$= \left(\dfrac{D+d}{2}\right)\pi bc$

Cylinder

cylindrical surface $= \pi Dh$

total surface $= 2\pi R(R+h)$

volume $(V) = \pi R^2 h$

$= \dfrac{c^2 h}{4\pi}$

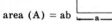

Parallelogram

area $(A) = ah$

Trapezoid

area $(A) = \dfrac{h}{2}(a + b)$

Trapezium

area $(A) = \tfrac{1}{2}\,[b(H+h) + ah + cH]$

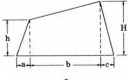

Regular Pentagon

area $(A) = 1.720\ a^2$

Regular Hexagon

area $(A) = 2.598\ a^2$

Regular Octagon

area $(A) = 4.828\ a^2$

Circle

circumference $(C) = 2\pi R$

$\qquad\qquad\qquad = \pi D$

area $(A) = \pi R^2$

Wire Gauge Standards

			Decimal parts of an inch				
Wire gauge no.	American or Brown & Sharpe	Birmingham or Stubs wire	Washburn & Moen on steel wire gauge	American S. & W. Co.'s music wire	Imperial wire gauge	Stubs steel wire	U.S. standard for plate
00000	0.516549	0.500	0.4305	0.005	4.432	....	0.43775
0000	0.460	0.454	0.3938	0.006	0.400	....	0.40625
000	0.40964	0.425	0.3625	0.007	0.372	....	0.375
00	0.3648	0.380	0.3310	0.008	0.348	....	0.34375
0	0.32486	0.340	0.3065	0.009	0.324	....	0.3125
1	0.2893	0.300	0.2830	0.010	0.300	0.227	0.28125
2	0.25763	0.284	0.2625	0.011	0.276	0.219	0.265625
3	0.22942	0.259	0.2437	0.012	0.252	0.212	0.250
4	0.20431	0.238	0.2253	0.013	0.232	0.207	0.234375
5	0.18194	0.220	0.2070	0.014	0.212	0.204	0.21875
6	0.16202	0.203	0.1920	0.016	0.192	0.201	0.203125
7	0.14428	0.180	0.1770	0.018	0.176	0.199	0.1875
8	0.12849	0.165	0.1620	0.020	0.160	0.197	0.171875
9	0.11443	0.148	0.1483	0.022	0.144	0.194	0.15625
10	0.10189	0.134	0.1350	0.024	0.128	0.191	0.140625
11	0.090742	0.120	0.1205	0.026	0.116	0.188	0.125
12	0.080808	0.109	0.1055	0.029	0.104	0.185	0.109375
13	0.071961	0.095	0.0915	0.031	0.092	0.182	0.09375
14	0.064084	0.083	0.0800	0.033	0.080	0.180	0.078125
15	0.057068	0.072	0.0720	0.035	0.072	0.178	0.0703125
16	0.05082	0.065	0.0625	0.037	0.064	0.175	0.0625
17	0.045257	0.058	0.0540	0.039	0.056	0.172	0.05625
18	0.040303	0.049	0.0475	0.041	0.048	0.168	0.050
19	0.03589	0.042	0.0410	0.043	0.040	0.164	0.04375
20	0.031961	0.035	0.0348	0.045	0.036	0.161	0.0375
21	0.028462	0.032	0.0317	0.047	0.032	0.157	0.034375
22	0.025347	0.028	0.0286	0.049	0.028	0.155	0.03125
23	0.022571	0.025	0.0258	0.051	0.024	0.153	0.028125
24	0.0201	0.022	0.0230	0.055	0.022	0.151	0.025
25	0.0179	0.020	0.0204	0.059	0.020	0.148	0.021875
26	0.01594	0.018	0.0181	0.063	0.018	0.146	0.01875
27	0.014195	0.016	0.0173	0.067	0.0164	0.143	0.0171875
28	0.012641	0.014	0.0162	0.071	0.0149	0.139	0.015625
29	0.011257	0.013	0.0150	0.075	0.0136	0.134	0.0140625
30	0.010025	0.012	0.0140	0.080	0.0124	0.127	0.0125

Wire Gauge Standards (Cont'd)

Wire gauge no.	Decimal parts of an inch							
	American or Brown & Sharpe	Birmingham or Stubs wire	Washburn & Moen on steel wire gauge	American S. & W. Co.'s music wire	Imperial wire gauge	Stubs steel wire	U.S. standard for plate
31	0.008928	0.010	0.0132	0.085	0.0116	0.120	0.0109375
32	0.00795	0.009	0.0128	0.090	0.0108	0.115	0.01015625
33	0.00708	0.008	0.0118	0.095	0.0100	0.112	0.009375
34	0.006304	0.007	0.0104	....	0.0092	0.110	0.00859375
35	0.005614	0.005	0.0095	....	0.0084	0.108	0.0078125
36	0.005	0.004	0.0090	....	0.0076	0.106	0.00703125
37	0.004453	....	0.0085	....	0.0068	0.103	0.006640625
38	0.003965	....	0.0080	....	0.0060	0.101	0.00625
39	0.003531	....	0.0075	....	0.0052	0.099	
40	0.003144	....	0.0070	....	0.0048	0.097	

Metal Weights

Material	Chemical Symbol	Weight, in Pounds Per Cubic Inch	Weight, in Pounds Per Cubic Foot
Aluminum	Al	.093	160
Antimony	Sb	.2422	418
Brass	—	.303	524
Bronze	—	.320	552
Chromium	Cr	.2348	406
Copper	Cu	.323	450
Gold	Au	.6975	1205
Iron (cast)	Fe	.260	450
Iron (wrought)	Fe	.2834	490
Lead	Pb	.4105	710
Manganese	Mn	.2679	463
Mercury	Hg	.491	849
Molybdenum	Mo	.309	534
Monel	—	.318	550
Platinum	Pt	.818	1413
Steel (mild)	Fe	.2816	490
Steel (stainless)	—	.277	484
Tin	Sn	.265	459
Titanium	Ti	.1278	221
Zinc	Zn	.258	446

Functions of Numbers

No.	Square	Cube	Square Root	Cubic Root	Logarithm	1000 x Reciprocal	No. = Diameter	
							Circum.	Area
1	1	1	1.0000	1.0000	0.00000	1000.000	3.142	0.7854
2	4	8	1.4142	1.2599	0.30103	500.000	6.283	3.1416
3	9	27	1.7321	1.4422	0.47712	333.333	9.425	7.0686
4	16	64	2.0000	1.5874	0.60206	250.000	12.566	12.5664
5	25	125	2.2361	1.7100	0.69897	200.000	15.708	19.6350
6	36	216	2.4495	1.8171	0.77815	166.667	18.850	28.2743
7	49	343	2.6458	1.9129	0.84510	142.857	21.991	38.4845
8	64	512	2.8284	2.0000	0.90309	125.000	25.133	50.2655
9	81	729	3.0000	2.0801	0.95424	111.111	28.274	63.6173
10	100	1000	3.1623	2.1544	1.00000	100.000	31.416	78.5398
11	121	1331	3.3166	2.2240	1.04139	90.9091	34.558	95.0332
12	144	1728	3.4641	2.2894	1.07918	83.3333	37.699	113.097
13	169	2197	3.6056	2.3513	1.11394	76.9231	40.841	132.732
14	196	2744	3.7417	2.4101	1.14613	71.4286	43.982	153.938
15	225	3375	3.8730	2.4662	1.17609	66.6667	47.124	176.715
16	256	4096	4.0000	2.5198	1.20412	62.5000	50.265	201.062
17	289	4913	4.1231	2.5713	1.23045	58.8235	53.407	226.980
18	324	5832	4.2426	2.6207	1.25527	55.5556	56.549	254.469
19	361	6859	4.3589	2.6684	1.27875	52.6316	59.690	283.529
20	400	8000	4.4721	2.7144	1.30103	50.0000	62.832	314.159
21	441	9261	4.5826	2.7589	1.32222	47.6190	65.973	346.361
22	484	10648	4.6904	2.8020	1.34242	45.4545	69.115	380.133
23	529	12167	4.7958	2.8439	1.36173	43.4783	72.257	415.476
24	576	13824	4.8990	2.8845	1.38021	41.6667	75.398	452.389
25	625	15625	5.0000	2.9240	1.39794	40.0000	78.540	490.874
26	676	17576	5.0990	2.9625	1.41497	38.4615	81.681	530.929
27	729	19683	5.1962	3.0000	1.43136	37.0370	84.823	572.555
28	784	21952	5.2915	3.0366	1.44716	35.7143	87.965	615.752
29	841	24389	5.3852	3.0723	1.46240	34.4828	91.106	660.520
30	900	27000	5.4772	3.1072	1.47712	33.3333	94.248	706.858
31	961	29791	5.5678	3.1414	1.49136	32.2581	97.389	754.768
32	1024	32768	5.6569	3.1748	1.50515	31.2500	100.531	804.248
33	1089	35937	5.7446	3.2075	1.51851	30.3030	103.673	855.299
34	1156	39304	5.8310	3.2396	1.53148	29.4118	106.814	907.920
35	1225	42875	5.9161	3.2711	1.54407	28.5714	109.956	962.113
36	1296	46656	6.0000	3.3019	1.55630	27.7778	113.097	1017.88
37	1369	50653	6.0828	3.3322	1.56820	27.0270	116.239	1075.21
38	1444	54872	6.1644	3.3620	1.57978	26.3158	119.381	1134.11
39	1521	59319	6.2450	3.3912	1.59106	25.6410	122.522	1194.59
40	1600	64000	6.3246	3.4200	1.60206	25.0000	125.66	1256.64
41	1681	68921	6.4031	3.4482	1.61278	24.3902	128.81	1320.25
42	1764	74088	6.4807	3.4760	1.62325	23.8095	131.95	1385.44
43	1849	79507	6.5574	3.5034	1.63347	23.2558	135.09	1452.20
44	1936	85184	6.6332	3.5303	1.64345	22.7273	138.23	1520.53
45	2025	91125	6.7082	3.5569	1.65321	22.2222	141.37	1590.43
46	2116	97336	6.7823	3.5830	1.66276	21.7391	144.51	1661.90
47	2209	103823	6.8557	3.6088	1.67210	21.2766	147.65	1734.94

Functions of Numbers (Cont'd)

No.	Square	Cube	Square Root	Cubic Root	Logarithm	1000 x Reciprocal	No. = Diameter Circum.	Area
48	2304	110592	6.9282	3.6342	1.68124	20.8333	150.80	1809.56
49	2401	117649	7.0000	3.6593	1.69020	20.4082	153.94	1885.74
50	2500	125000	7.0711	3.6840	1.69897	20.0000	157.08	1963.50
51	2601	132651	7.1414	3.7084	1.70757	19.6078	160.22	2042.82
52	2704	140608	7.2111	3.7325	1.71600	19.2308	163.36	2123.72
53	2809	148877	7.2801	3.7563	1.72428	18.8679	166.50	2206.18
54	2916	157464	7.3485	3.7798	1.73239	18.5185	169.65	2290.22
55	3025	166375	7.4162	3.8030	1.74036	18.1818	172.79	2375.83
56	3136	175616	7.4833	3.8259	1.74819	17.8571	175.93	2463.01
57	3249	185193	7.5498	3.8485	1.75587	17.5439	179.07	2551.76
58	3364	195112	7.6158	3.8709	1.76343	17.2414	182.21	2642.08
59	3481	205379	7.6811	3.8930	1.77085	16.9492	185.35	2733.97
60	3600	216000	7.7460	3.9149	1.77815	16.6667	188.50	2827.43
61	3721	226981	7.8102	3.9365	1.78533	16.3934	191.64	2922.47
62	3844	238328	7.8740	3.9579	1.79239	16.1290	194.78	3019.07
63	3969	250047	7.9373	3.9791	1.79934	15.8730	197.92	3117.25
64	4096	262144	8.0000	4.0000	1.80618	15.6250	201.06	3216.99
65	4225	274625	8.0623	4.0207	1.81291	15.3846	204.20	3318.31
66	4356	287496	8.1240	4.0412	1.81954	15.1515	207.35	3421.19
67	4489	300763	8.1854	4.0615	1.82607	14.9254	210.49	3525.65
68	4624	314432	8.2462	4.0817	1.83251	14.7059	213.63	3631.68
69	4761	328509	8.3066	4.1016	1.83885	14.4928	216.77	3739.28
70	4900	343000	8.3666	4.1213	1.84510	14.2857	219.91	3848.45
71	5041	357911	8.4261	4.1408	1.85126	14.0845	223.05	3959.19
72	5184	373248	8.4853	4.1602	1.85733	13.8889	226.19	4071.50
73	5329	389017	8.5440	4.1793	1.86332	13.6986	229.34	4185.39
74	5476	405224	8.6023	4.1983	1.86923	13.5135	232.48	4300.84
75	5625	421875	8.6603	4.2172	1.87506	13.3333	235.62	4417.86
76	5776	438976	8.7178	4.2358	1.88081	13.1579	238.76	4536.46
77	5929	456533	8.7750	4.2543	1.88649	12.9870	241.90	4656.63
78	6084	474552	8.8318	4.2727	1.89209	12.8205	245.04	4778.36
79	6241	493039	8.8882	4.2908	1.89763	12.6582	248.19	4901.67
80	6400	512000	8.9443	4.3089	1.90309	12.5000	251.33	5026.55
81	6561	531441	9.0000	4.3267	1.90849	12.3457	254.47	5153.00
82	6724	551368	9.0554	4.3445	1.91381	12.1951	257.61	5281.02
83	6889	571787	9.1104	4.3621	1.91908	12.0482	260.75	5410.61
84	7056	592704	9.1652	4.3795	1.92428	11.9048	263.89	5541.77
85	7225	614125	9.2195	4.3968	1.92942	11.7647	267.04	5674.50
86	7396	636056	9.2736	4.4140	1.93450	11.6279	270.18	5808.80
87	7569	658503	9.3274	4.4310	1.93952	11.4943	273.32	5944.68
88	7744	681472	9.3808	4.4480	1.94448	11.3636	276.46	6082.12
89	7921	704969	9.4340	4.4647	1.94939	11.2360	279.60	6221.14
90	8100	729000	9.4868	4.4814	1.95424	11.1111	282.74	6361.73
91	8281	753571	9.5394	4.4979	1.95904	10.9890	285.88	6503.88
92	8464	778688	9.5917	4.5144	1.96379	10.8696	289.03	6647.61
93	8649	804357	9.6437	4.5307	1.96848	10.7527	292.17	6792.91

Functions of Numbers (Cont'd)

No.	Square	Cube	Square Root	Cubic Root	Logarithm	1000 x Reciprocal	No. = Diameter Circum.	Area
94	8836	830584	9.6954	4.5468	1.97313	10.6383	295.31	6939.78
95	9025	857375	9.7468	4.5629	1.97772	10.5263	298.45	7088.22
96	9216	884736	9.7980	4.5789	1.98227	10.4167	301.59	7238.23
97	9409	912673	9.8489	4.5947	1.98677	10.3093	304.73	7389.81
98	9604	941192	9.8995	4.6104	1.99123	10.2041	307.88	7542.96
99	9801	970299	9.9499	4.6261	1.99564	10.1010	311.02	7697.69

Metric and English Equivalent Measures

MEASURES OF LENGTH

Metric		English
1 meter	=	39.37 inches, or 3.28083 feet, or 1.09361 yards
.3048 meter	=	1 foot
1 centimeter	=	.3937 inch
2.54 centimeters	=	1 inch
1 millimeter	=	.03937 inch, or nearly 1-25 inch
25.4 millimeters	=	1 inch
1 kilometer	=	1093.61 yards, or 0.62137 mile

MEASURES OF WEIGHT

Metric		English
1 gram	=	15.432 grains
.0648 gram	=	1 grain
28.35 grams	=	1 ounce avoirdupois
1 kilogram	=	2.2046 pounds
.4536 kilogram	=	1 pound
1 metric ton 1000 kilograms	=	$\begin{cases} .9842 \text{ ton of 2240 pounds} \\ 19.68 \text{ cwt.} \\ 2204.6 \text{ pounds} \end{cases}$
1.016 metric tons 1016 kilograms	=	1 ton of 2240 pounds

MEASURES OF CAPACITY

Metric		English
1 liter (= 1 cubic decimeter)	=	$\begin{cases} 61.023 \text{ cubic inches} \\ .03531 \text{ cubic foot} \\ .2642 \text{ gal. (American)} \\ 2.202 \text{ lbs. of water at } 62° \text{ F.} \end{cases}$
28.317 liters	=	1 cubic foot
3.785 liters	=	1 gallon (American)
4.543 liters	=	1 gallon (Imperial)

English Conversion Table

Length

Inches	×	.0833	= feet
Inches	×	.02778	= yards
Inches	×	.00001578	= miles
Feet	×	.3333	= yards
Feet	×	.0001894	= miles
ards	×	36.00	= inches
Yards	×	3.00	= feet
Yards	×	.0005681	= miles
Miles	×	63360.00	= inches
Miles	×	5280.00	= feet
Miles	×	1760.00	= yards
Circumference of circle	×	.3188	= diameter
Diameter of circle	×	3.1416	= circumference

Area

Square inches	×	.00694	= square feet
Square inches	×	.0007716	= square yards
Square feet	×	144.00	= square inches
Square feet	×	.11111	= square yards
Square yards	×	1296.00	= square inches
Square yards	×	9.00	= square feet
Dia. of circle squared	×	.7854	= area
Dia. of sphere squared	×	3.1416	= surface

Volume

Cubic inches	×	.0005787	= cubic feet
Cubic inches	×	.00002143	= cubic yards
Cubic inches	×	.004329	= U. S. gallons
Cubic feet	×	1728.00	= cubic inches
Cubic feet	×	.03704	= cubic yards
Cubic feet	×	7.4805	= U. S. gallons
Cubic yards	×	46656.00	= cubic inches
Cubic yards	×	27.00	= cubic feet
Dia. of sphere cubed	×	.5236	= volume

Weight

Grains (avoirdupois)	×	.002286	= ounces
Ounces (avoirdupois)	×	.0625	= pounds
Ounces (avoirdupois)	×	.00003125	= tons
Pounds (avoirdupois)	×	16.00	= ounces
Pounds (avoirdupois)	×	.01	= hundredweight
Pounds (avoirdupois)	×	.0005	= tons
Tons (avoirdupois)	×	32000.00	= ounces
Tons (avoirdupois)	×	2000.00	= pounds

English Conversion Table

Energy

Horsepower	×	33000.	= ft.-lbs. per min.
B. t. u.	×	778.26	= ft.-lbs.
Ton of refrigeration	×	200.	= B. t. u. per min.

Pressure

Lbs. per sq. in.	×	2.31	= ft. of water (60°F.)
Ft. of water (60°F.)	×	.433	= lbs. per sq. in.
Ins. of water (60°F.)	×	.0361	= lbs. per sq. in.
Lbs. per sq. in.	×	27.70	= ins. of water (60°F.)
Lbs. per sq. in.	×	2.041	= ins. of Hg. (60°F.)
Ins. of Hg (60°F.)	×	.490	= lbs. per sq. in.

Power

Horsepower	×	746.	= watts
Watts	×	.001341	= horsepower
Horsepower	×	42.4	= B. t. u. per min.

Water Factors (at point of greatest density—39.2°F.)

Miners inch (of water)	×	8.976	= U. S. gals. per min.
Cubic inches (of water)	×	.57798	= ounces
Cubic inches (of water)	×	.036124	= pounds
Cubic inches (of water)	×	.004329	= U. S. gallons
Cubic inches (of water)	×	.003607	= English gallons
Cubic feet (of water)	×	62.425	= pounds
Cubic feet (of water)	×	.03121	= tons
Cubic feet (of water)	×	7.4805	= U. S. gallons
Cubic inches (of water)	×	6.232	= English gallons
Cubic foot of ice	×	57.2	= pounds
Ounces (of water)	×	1.73	= cubic inches
Pounds (of water)	×	26.68	= cubic inches
Pounds (of water)	×	.01602	= cubic feet
Pounds (of water)	×	.1198	= U. S. gallons
Pounds (of water)	×	.0998	= English gallons
Tons (of water)	×	32.04	= cubic feet
Tons (of water)	×	239.6	= U. S. gallons
Tons (of water)	×	199.6	= English gallons
U. S. gallons	×	231.00	= cubic inches
U. S. gallons	×	.13368	= cubic feet
U. S. gallons	×	8.345	= pounds
U. S. gallons	×	.8327	= English gallons
U. S. gallons	×	3.785	= liters
English gallons (Imperial)	×	277.41	= cubic inches
English gallons (Imperial)	×	.1605	= cubic feet

English gallons (Imperial)	×	10.02	= pounds
English gallons (Imperial)	×	1.201	= U. S. gallons
English gallons (Imperial)	×	4.546	= liters

Metric Conversion Table

Length

Millimeters	×	.03937	= inches
Millimeters	÷	25.4	= inches
Centimeters	×	.3937	= inches
Centimeters	÷	2.54	= inches
Meters	×	39.37	= inches (Act. Cong.)
Meters	×	3.281	= feet
Meters	×	1.0936	= yards
Kilometers	×	.6214	= miles
Kilometers	÷	1.6093	= miles
Kilometers	×	3280.8	= feet

Area

Sq. Millimeters	×	.00155	= sq. in.
Sq. Millimeters	÷	645.2	= sq. in.
Sq. Centimeters	×	.155	= sq. in.
Sq. Centimeters	÷	6.452	= sq. in.
Sq. Meters	×	10.764	= sq. ft.
Sq. Kilometers	×	247.1	= acres
Hectares	×	2.471	= acres

Volume

Cu. Centimeters	÷	16.387	= cu. in.
Cu. Centimeters	÷	3.69	= fl. drs. (U.S.P.)
Cu. Centimeters	÷	29.57	= fl. oz. (U.S.P.)
Cu. Meters	×	35.314	= cu. ft.
Cu. Meters	×	1.308	= cu. yards
Cu. Meters	×	264.2	= gals. (231 cu. in.)
Litres	×	61.023	= cu. in. (Act. Cong.)
Litres	×	33.82	= fl. oz. (U.S.J.)
Litres	×	.2642	= gals. (231 cu. in.)
Litres	÷	3.785	= gals. (231 cu. in.)
Litres	÷	28.317	= cu. ft.
Hectolitres	×	3.531	= cu. ft.
Hectolitres	×	2.838	= bu. (2150.42 cu. in.)
Hectolitres	×	.1308	= cu. yds.
Hectolitres	×	26.42	= gals. (231 cu. in.)

Weight

| Grams | × | 15.432 | = grains (Act. Cong.) |
| Grams | ÷ | 981. | = dynes |

Grams (water)	÷	29.57	= fl. oz.
Grams	÷	28.35	= oz. avoirdupois
Kilo-grams	×	2.2046	= lbs.

Weight

Kilo-grams	×	35.27	= oz. avoirdupois
Kilo-grams	×	.0011023	= tons (2000 lbs.)
Tonneau (Metric ton)	×	1.1023	= tons (2000 lbs.)
Tonneau (Metric ton)	×	2204.6	= lbs.

Unit Weight

Grams per cu. cent.	÷	27.68	= lbs. per cu. in.
Kilo per meter	×	.672	= lbs. per ft.
Kilo per cu. meter	×	.06243	= lbs. per cu. ft.
Kilo per Cheval	×	2.235	= lbs. per h. p.
Grams per liter	×	.06243	= lbs. per cu. ft.

Pressure

Kilo-grams per sq. cm.	×	14.223	= lbs. per sq. in.
Kilo-grams per sq. cm.	×	32.843	= ft. of water (60°F.)
Atmospheres (international)	×	14.696	= lbs. per sq. in.

Energy

Joule	×	.7376	= ft. lbs.
Kilo-gram meters	×	7.233	= ft. lbs.

Power

Cheval vapeur	×	.9863	= h. p.
Kilo-watts	×	1.341	= h. p.
Watts	÷	746.	= h. p.
Watts	×	.7373	= ft. lbs. per sec

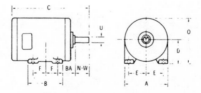

USE WITH TABLES ON PAGES 232 AND 233

U Nema Motor Frame Dimensions

Horsepower Rating			U Frame	U	Shaft Keyseat		Key Length	N-W	A Max.	B Max.	C	D	E	F	BA	O
3600	1800	1200	No.		Width	Depth										
1 1/2	1	3/4	182	7/8	3/16	3/32	1 3/8	2 1/4	8 7/8	6 1/2	12 1/8	4 1/2	3 3/4	2 1/4	2 3/4	8 15/16
2 & 3	1 1/2 & 2	1 & 1 1/2	184	7/8	3/16	3/32	1 3/8	2 1/4	8 7/8	7 1/2	13 1/8	4 1/2	3 3/4	2 3/4	2 3/4	8 15/16
5	3	2	213	1 1/8	1/4	1/8	2	3	10 3/8	7 1/2	15 5/16	5 1/4	4 1/4	2 3/4	3 1/2	10 7/16
7 1/2	5	3	215	1 1/8	1/4	1/8	2	3	10 3/8	9	16 13/16	5 1/4	4 1/4	3 1/2	3 1/2	10 7/16
10	7 1/2	5	254U	1 3/8	5/16	5/32	2 3/4	3 3/4	12 7/16	10 3/4	20 1/4	6 1/4	5	4 1/8	4 1/4	12 1/2
15	10	7 1/2	256U	1 3/8	5/16	5/32	2 3/4	3 3/4	12 7/16	12 1/2	22	6 1/4	5	5	4 1/4	12 1/2
20	15	10	284U	1 5/8	3/8	3/16	3 3/4	4 7/8	13 7/8	12 1/2	23 11/16	7	5 1/2	4 3/4	4 3/4	13 15/16
25	20	......	286U	1 5/8	3/8	3/16	3 3/4	4 7/8	13 7/8	12 1/2	25 3/16	7	5 1/2	5 1/2	4 3/4	13 15/16
......	25	15	324U	1 7/8	1/2	1/4	4 1/4	5 5/8	15 7/8	14	26 3/8	8	6 1/4	5 1/4	5 1/4	15 15/16
30	......	......	324S	1 5/8	3/8	3/16	1 7/8	3 1/4	15 7/8	14	24	8	6 1/4	5 1/4	5 1/4	15 15/16
......	30	20	326U	1 7/8	1/2	1/4	4 1/4	5 5/8	15 7/8	14	27 7/8	8	6 1/4	6	5 1/4	15 15/16
40	......	......	326S	1 5/8	3/8	3/16	1 7/8	3 1/4	15 7/8	14	25 1/2	8	6 1/4	6	5 1/4	15 15/16
......	40	25	364U	2 1/8	1/2	1/4	5	6 3/8	17 5/8	15 1/4	29 3/16	9	7	5 5/8	5 7/8	17 13/16
50	......	......	364US	1 7/8	1/2	1/4	2	3 3/4	17 5/8	15 1/4	26 9/16	9	7	5 5/8	5 7/8	17 13/16
......	50	30	365U	2 1/8	1/2	1/4	5	6 3/8	17 5/8	16 1/4	30 3/16	9	7	6 1/8	5 7/8	17 13/16
60	......	......	365US	1 7/8	1/2	1/4	2	3 3/4	19 3/4	16 1/4	27 9/16	9	7	6 1/8	5 7/8	17 13/16
......	60	40	404U	2 3/8	5/8	5/16	5 1/2	7 1/8	19 3/4	16 1/4	32 7/16	10	8	6 1/8	6 5/8	19 7/8
75	......	......	404US	2 1/8	1/2	1/4	2 3/4	4 1/4	19 3/4	16 1/4	39 9/16	10	8	6 1/8	6 5/8	19 7/8
......	75	50	405U	2 3/8	5/8	5/16	5 1/2	7 1/8	19 3/4	17 1/4	33 15/16	10	8	6 7/8	6 5/8	19 7/8
100	......	......	405US	2 1/8	1/2	1/4	2 3/4	4 1/4	19 3/4	17 1/4	31 1/16	10	8	6 7/8	6 5/8	19 7/8

T Nema Motor Frame Dimensions

Horsepower Rating 3600	Horsepower Rating 1800	Horsepower Rating 1200	T Frame No.	U	Shaft Keyseat Width	Shaft Keyseat Depth	Key Length	N-W	A Max.	B Max.	C	D	E	F	BA	O
1½	1	¾	143T	⅞	3/16	3/32	1⅜	2¼	7	6	12 ⅝	3½	2¾	2	2¼	7
2 & 3	1½ & 2	1	145T	⅞	3/16	3/32	1⅜	2¼	7	6	12 ⅝	3½	2¾	2½	2¼	7
5	3	1½	182T	1⅛	¼	⅛	1¾	2¾	9	6½	12 ¾	4½	3¾	2¼	2¾	9
7½	5	2	184T	1⅛	¼	⅛	1¾	2¾	9	7½	13 ¾	4½	3¾	2¾	2¾	9
10	7½	3	213T	1⅜	5/16	5/32	2⅜	3⅜	10½	7½	15 13/16	5¼	4¼	2¾	3½	10½
15	10	5	215T	1⅜	5/16	5/32	2⅜	3⅜	10½	9	17 5/16	5¼	4¼	3½	3½	10½
20	15	7½	254T	1⅝	⅜	3/16	2⅞	4	12½	10¾	20 ½	6¼	5	4⅛	4¼	12½
25	20	10	256T	1⅝	⅜	3/16	2⅞	4	12½	12½	22 ¼	6¼	5	5	4¼	12½
......	25	15	284T	1⅞	½	¼	3¼	4⅝	14	12½	23 5/16	7	5½	4¾	4¾	14
30	25	15	284TS	1⅝	⅜	3/16	1⅞	3¼	14	12½	22	7	5½	4¾	4¾	14
......	30	20	286T	1⅞	½	¼	3¼	4⅝	14	14	24 ⅞	7	5½	5½	4¾	14
40	30	20	286TS	1⅝	⅜	3/16	1⅞	3¼	14	14	23 ½	7	5½	5½	4¾	14
......	40	25	324T	2⅛	½	¼	3⅞	5¼	16	14	26 ½	8	6¼	5¼	5¼	16
50	40	25	324TS	1⅞	½	¼	2	3¾	16	14	24 ⅝	8	6¼	5¼	5¼	16
......	50	30	326T	2⅛	½	¼	3⅞	5¼	16	15½	27 ¾	8	6¼	6	5¼	16
60	50	30	326TS	1⅞	½	¼	2	3¾	16	15½	26 ⅛	8	6¼	6	5¼	16
......	60	40	364T	2⅜	⅝	5/16	4¼	5⅞	18	15¼	28 ¾	9	7	5⅝	5⅞	18
75	60	......	364TS	1⅞	½	¼	2	3¾	18	15¼	26 9/16	9	7	5⅝	5⅞	18
......	75	50	365T	2⅜	⅝	5/16	4¼	5⅞	18	16¼	29 ¾	9	7	6⅛	5⅞	18
100	75	......	365TS	1⅞	½	¼	2	3¾	18	16¼	27 9/16	9	7	6⅛	5⅞	18

Taper Pins

All sizes have a taper of 0.250 per foot

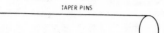

TAPER PINS

Size no. of pin	Length of pin	Large end of pin	Small end of reamer	Drill size for reamer
0	1	0.156	0.135	28
1	1 1/4	0.172	0.146	25
2	1 1/2	0.193	0.162	19
3	1 3/4	0.219	0.183	12
4	2	0.250	0.208	3
5	2 1/4	0.289	0.242	1/4
6	3 1/4	0.341	0.279	9/32
7	3 3/4	0.409	0.331	11/32
8	4 1/2	0.492	0.398	13/32
9	5 1/4	0.591	0.482	31/64
10	6	0.706	0.581	19/32
11	7 1/4	0.857	0.706	23/32
12	8 3/4	1.013	0.842	55/64

Keyway Data

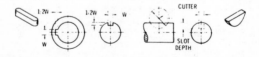

Shaft dia.	Square keyways	Woodruff keyways*			
		Key no.	Thickness	Cutter dia.	Slot depth
0.500	1/8 × 1/16	404	0.1250	0.500	0.1405
0.562	1/8 × 1/16	404	0.1250	0.500	0.1405
0.625	5/32 × 5/64	505	0.1562	0.625	0.1669
0.688	3/16 × 3/32	606	0.1875	0.750	0.2193
0.750	3/16 × 3/32	606	0.1875	0.750	0.2193
0.812	3/16 × 3/32	606	0.1875	0.750	0.2193

Keyway Data (Cont'd)

Shaft dia.	Square keyways	Woodruff keyways*			
		Key no.	Thickness	Cutter dia.	slot depth
0.875	$\frac{7}{32} \times \frac{7}{64}$	607	0.1875	0.875	0.2763
0.938	$\frac{1}{4} \times \frac{1}{8}$	807	0.2500	0.875	0.2500
1.000	$\frac{1}{4} \times \frac{1}{8}$	808	0.2500	1.000	0.3130
1.125	$\frac{5}{16} \times \frac{5}{32}$	1009	0.3125	1.125	0.3228
1.250	$\frac{5}{16} \times \frac{5}{32}$	1010	0.3125	1.250	0.3858
1.375	$\frac{3}{8} \times \frac{3}{16}$	1210	0.3750	1.250	0.3595
1.500	$\frac{3}{8} \times \frac{3}{16}$	1212	0.3750	1.500	0.4535
1.625	$\frac{3}{8} \times \frac{3}{16}$	1212	0.3750	1.500	0.4535
1.750	$\frac{7}{16} \times \frac{7}{32}$				
1.875	$\frac{1}{2} \times \frac{1}{4}$				
2.000	$\frac{1}{2} \times \frac{1}{4}$				
2.250	$\frac{5}{8} \times \frac{5}{16}$				
2.500	$\frac{5}{8} \times \frac{5}{16}$				
2.750	$\frac{3}{4} \times \frac{3}{8}$				
3.000	$\frac{3}{4} \times \frac{3}{8}$				
3.250	$\frac{3}{4} \times \frac{3}{8}$				
3.500	$\frac{7}{8} \times \frac{7}{16}$				
4.000	$1 \times \frac{1}{2}$				

* The depth of a Woodruff Keyway is measured from the edge of the slot.

Dimensions of Standard Gib-head Keys, Square and Flat

Approved by ASA*

Diameters of shafts	Square type					Flat type				
	Key		Gib head			Key		Gib head		
	W	H	C	D	E	W	H	C	D	E
$\frac{1}{2}$- $\frac{9}{16}$	$\frac{1}{8}$	$\frac{1}{8}$	$\frac{1}{4}$	$\frac{7}{32}$	$\frac{5}{32}$	$\frac{1}{8}$	$\frac{3}{32}$	$\frac{3}{16}$	$\frac{1}{8}$	$\frac{1}{8}$
$\frac{5}{8}$- $\frac{7}{8}$	$\frac{3}{16}$	$\frac{3}{16}$	$\frac{5}{16}$	$\frac{9}{32}$	$\frac{7}{32}$	$\frac{3}{16}$	$\frac{1}{8}$	$\frac{1}{4}$	$\frac{3}{16}$	$\frac{5}{32}$
$\frac{15}{16}$-1 $\frac{1}{4}$	$\frac{1}{4}$	$\frac{1}{4}$	$\frac{3}{16}$	$\frac{11}{32}$	$\frac{11}{32}$	$\frac{1}{4}$	$\frac{3}{16}$	$\frac{5}{16}$	$\frac{1}{4}$	$\frac{3}{16}$

Dimensions of Standard Gib-head Keys, Square and Flat (Cont'd)

Approved by ASA*

Diameters of shafts	Square type					Flat type				
	Key		Gib head			Key		Gib head		
	W	H	C	D	E	W	H	C	D	E
1 5/16-1 3/8	5/16	5/16	9/16	13/32	13/32	5/16	1/4	3/8	5/16	1/4
1 7/16-1 3/4	3/8	3/8	11/16	15/32	15/32	3/8	1/4	7/16	3/8	5/16
1 13/16-2 1/4	1/2	1/2	7/8	19/32	5/8	1/2	3/8	5/8	1/2	7/16
2 5/16-2 3/4	5/8	5/8	1 1/16	23/32	3/4	5/8	7/16	3/4	5/8	1/2
2 7/8-3 1/4	3/4	3/4	1 1/4	7/8	7/8	3/4	1/2	7/8	3/4	5/8
3 3/8-3 3/4	7/8	7/8	1 1/2	1	1	7/8	5/8	1 1/16	7/8	3/4
3 7/8-4 1/2	1	1	1 3/4	1 3/16	1 3/16	1	3/4	1 1/4	1	13/16
4 3/4-5 1/2	1 1/4	1 1/4	2	1 7/16	1 7/16	1 1/4	7/8	1 1/2	1 1/4	1
5 3/4-6	1 1/2	1 1/2	2 1/2	1 3/4	1 3/4	1 1/2	1	1 3/4	1 1/2	1 1/4

* ASA B17.1—1934. Dimensions in inches.

BOLTING DIMENSIONS FOR 150 LB. FLANGES

Nom. Pipe Size	150 LB. STEEL FLANGES				Bolt Length For 125 Lb. Cast Iron Flanges
	Diam. Of Bolt Circle	Diam. Of Bolts	No. Of Bolts	Length Of Studs 1/16" Raised Face	
1/2	2 3/8	1/2	4	2 1/4	
3/4	2 3/4	1/2	4	2 1/4	
1	3 1/8	1/2	4	2 1/2	1 3/4
1 1/4	3 1/2	1/2	4	2 1/2	2
1 1/2	3 7/8	1/2	4	2 3/4	2
2	4 3/4	5/8	4	3	2 1/4
2 1/2	5 1/2	5/8	4	3 1/4	2 1/2
3	6	5/8	4	3 1/2	2 1/2
3 1/2	7	5/8	8	3 1/2	2 3/4
4	7 1/2	5/8	8	3 1/2	3
5	8 1/2	3/4	8	3 3/4	3
6	9 1/2	3/4	8	3 3/4	3 1/4
8	11 3/4	3/4	8	4	3 1/2
10	14 1/4	7/8	12	4 1/2	3 3/4
12	17	7/8	12	4 1/2	3 3/4

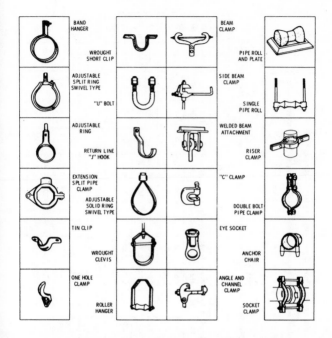

			TYPE OF WELD								
BEAD	FILLET	GROOVE					PLUG AND SLOT	FIELD WELD	WELD ALL AROUND	FLUSH
		SQUARE	V	BEVEL	U	J				

Labels within figure: BAND HANGER, WROUGHT SHORT CLIP, ADJUSTABLE SPLIT RING SWIVEL TYPE, "U" BOLT, ADJUSTABLE RING, RETURN LINE "J" HOOK, EXTENSION SPLIT PIPE CLAMP, ADJUSTABLE SOLID RING SWIVEL TYPE, TIN CLIP, WROUGHT CLEVIS, ONE HOLE CLAMP, ROLLER HANGER, BEAM CLAMP, PIPE ROLL AND PLATE, SIDE BEAM CLAMP, SINGLE PIPE ROLL, WELDED BEAM ATTACHMENT, RISER CLAMP, "C" CLAMP, DOUBLE BOLT PIPE CLAMP, EYE SOCKET, ANCHOR CHAIR, ANGLE AND CHANNEL CLAMP, SOCKET CLAMP

Weld symbol panels: ARROW (OR NEAR) SIDE OF JOINT, OTHER (OR FAR) SIDE OF JOINT, BOTH SIDES OF JOINT

Commercial Pipe Sizes and Wall Thicknesses

The following table lists the pipe sizes and wall thicknesses currently established as standard, or specifically:

1. The traditional standard weight, extra strong, and double extra strong pipe.
2. The pipe wall thickness schedules listed in American Standard B36.10, which are applicable to carbon steel and alloys other than stainless steels.
3. The pipe wall thickness schedules listed in American Standard B36.19, which are applicable only to stainless steels.

Nominal Pipe Size	Outside Diam.	Sched. 5*	Sched. 10*	Sched. 20	Sched. 30	Standard†	Sched. 40	Sched. 60	Extra Strong‡	Sched. 80	Sched. 100	Sched. 120	Sched. 140	Sched. 160	XX Strong
1/8	0.405	–	0.049	–	–	0.068	0.068	–	0.095	0.095	–	–	–	–	–
1/4	0.540	–	0.065	–	–	0.088	0.086	–	0.119	0.119	–	–	–	–	–
3/8	0.675	–	0.065	–	–	0.091	0.091	–	0.126	0.126	–	–	–	–	–
1/2	0.840	–	0.083	–	–	0.109	0.109	–	0.147	0.147	–	–	–	0.187	0.294
3/4	1.050	0.065	0.083	–	–	0.113	0.113	–	0.154	0.154	–	–	–	0.218	0.308
1	1.315	0.065	0.109	–	–	0.133	0.133	–	0.179	0.179	–	–	–	0.250	0.358
1 1/4	1.660	0.065	0.109	–	–	0.140	0.140	–	0.191	0.191	–	–	–	0.250	0.382
1 1/2	1.900	0.065	0.109	–	–	0.145	0.145	–	0.200	0.200	–	–	–	0.281	0.400
2	2.375	0.065	0.109	–	–	0.154	0.154	–	0.218	0.218	–	–	–	0.343	0.436
2 1/2	2.875	0.083	0.120	–	–	0.203	0.203	–	0.276	0.276	–	–	–	0.375	0.552
3	3.5	0.083	0.120	–	–	0.216	0.216	–	0.300	0.300	–	–	–	0.438	0.600
3 1/2	4.0	0.083	0.120	–	–	0.226	0.226	–	0.318	0.318	–	–	–	–	–

Commercial Pipe Sizes and Wall Thicknesses (Cont'd)

Nominal Pipe Size	Outside Diam.	Nominal Wall Thickness For													
		Sched. 5*	Sched. 10*	Sched. 20	Sched. 30	Standard†	Sched. 40	Sched. 60	Extra Strong‡	Sched. 80	Sched 100	Sched. 120	Sched. 140	Sched. 160	XX Strong
4	4.5	0.083	0.120	—	—	0.237	0.237	—	0.337	0.337	—	0.438	—	0.531	0.674
5	5.563	0.109	0.134	—	—	0.258	0.258	—	0.375	0.375	—	0.500	—	0.625	0.750
6	6.625	0.109	0.134	—	—	0.280	0.280	—	0.432	0.432	—	0.562	—	0.718	0.864
8	8.625	0.109	0.148	0.250	0.277	0.322	0.322	0.406	0.500	0.500	0.593	0.718	0.812	0.906	0.875
10	10.75	0.134	0.165	0.250	0.307	0.365	0.365	0.500	0.500	0.593	0.713	0.843	1.000	1.125	—
12	12.75	0.156	0.180	0.250	0.330	0.375	0.406	0.562	0.500	0.687	0.843	1.000	1.125	1.312	—
14 O.D.	14.0	—	0.250	0.312	0.375	0.375	0.438	0.593	0.500	0.750	0.937	1.093	1.250	1.406	—
16 O.D.	16.0	—	0.250	0.312	0.375	0.375	0.500	0.656	0.500	0.843	1.031	1.218	1.438	1.593	—
18 O.D.	18.0	—	0.250	0.312	0.438	0.375	0.562	0.750	0.500	0.937	1.156	1.375	1.562	1.781	—
20 O.D.	20.0	—	0.250	0.375	0.500	0.375	0.593	0.812	0.500	1.031	1.281	1.500	1.750	1.968	—
22 O.D.	22.0	—	0.250	0.250	—	0.375	—	—	0.500	—	—	—	—	—	—
24 O.D.	24.0	—	0.250	0.375	0.562	0.375	0.687	0.968	0.500	1.218	1.531	1.812	2.062	2.343	—
26 O.D.	26.0	—	—	—	—	0.375	—	—	0.500	—	—	—	—	—	—
30 O.D.	30.0	—	0.312	0.500	0.625	0.375	—	—	0.500	—	—	—	—	—	—
34 O.D.	34.0	—	—	—	—	0.375	—	—	0.500	—	—	—	—	—	—
36 O.D.	36.0	—	—	—	—	0.375	—	—	0.500	—	—	—	—	—	—
42 O.D.	42.0	—	—	—	—	0.375	—	—	0.500	—	—	—	—	—	—

BLIND FLANGES

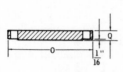

Nom. Pipe Size	150 LB.		300 LB.	
	Outside Diam. of Flange O	Thickness Q	Outside Diam. of Flange O	Thickness Q
1/2	3 1/2	7/16	3 3/4	9/16
3/4	3 7/8	1/2	4 5/8	5/8
1	4 1/4	9/16	4 7/8	11/16
1 1/4	4 5/8	5/8	5 1/4	3/4
1 1/2	5	11/16	6 1/8	13/16
2	6	3/4	6 1/2	7/8
2 1/2	7	7/8	7 1/2	1
3	7 1/2	15/16	8 1/4	1 1/8
3 1/2	8 1/2	15/16	9	1 3/16
4	9	15/16	10	1 1/4
5	10	15/16	11	1 3/8
6	11	1	12 1/2	1 7/16
8	13 1/2	1 1/8	15	1 5/8
10	16	1 3/16	17 1/2	1 7/8
12	19	1 1/4	20 1/2	2

WELDING NECK FLANGES

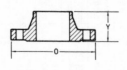

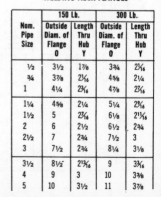

Nom. Pipe Size	150 Lb.		300 Lb.	
	Outside Diam. of Flange O	Length Thru Hub Y	Outside Diam. of Flange O	Length Thru Hub Y
1/2	3 1/2	1 7/8	3 3/4	2 1/16
3/4	3 7/8	2 1/16	4 5/8	2 1/4
1	4 1/4	2 3/16	4 7/8	2 7/16
1 1/4	4 5/8	2 1/4	5 1/4	2 9/16
1 1/2	5	2 7/16	6 1/8	2 11/16
2	6	2 1/2	6 1/2	2 3/4
2 1/2	7	2 3/4	7 1/2	3
3	7 1/2	2 3/4	8 1/4	3 1/8
3 1/2	8 1/2	2 13/16	9	3 3/16
4	9	3	10	3 3/8
5	10	3 1/2	11	3 7/8

WELDING NECK FLANGES (Cont'd)

Nom. Pipe Size	150 Lb.		300 Lb.	
	Outside Diam. of Flange O	Length Thru Hub Y	Outside Diam. of Flange O	Length Thru Hub Y
6	11	3½	12½	3⅞
8	13½	4	15	4⅜
10	16	4	17½	4⅝
12	19	4½	20½	5⅛
14	21	5	23	5⅝
16	23½	5	25½	5¾
18	25	5½	28	6¼

STANDARD CAST IRON COMPANION FLANGES AND BOLTS

(For working pressures up to 125 psi steam, 175 psi WOG)

Size In Inches	Diam. Of Flange, In Inches	Bolt Circle, In Inches	No. Of Bolts	Size Of Bolts, In Inches	Length Of Bolts, In Inches
¾	3½	2½	4	⅜	1⅜
1	4¼	3⅛	4	½	1½
1¼	4⅝	3½	4	½	1½
1½	5	3⅞	4	½	1¾
2	6	4¾	4	⅝	2
2½	7	5½	4	⅝	2¼
3	7½	6	4	⅝	2½
3½	8½	7	8	⅝	2½
4	9	7½	8	⅝	2¾
5	10	8½	8	¾	3
6	11	9½	8	¾	3
8	13½	11¾	8	¾	3¼
10	16	14¼	12	⅞	3½
12	19	17	12	⅞	3¾
14	21	18¾	12	1	4¼
16	23½	21¼	16	1	4¼

EXTRA HEAVY CAST IRON COMPANION FLANGES AND BOLTS

(For working pressures up to 250 psi steam, 400 psi WOG)

Pipe Size Inches	Diam. Of Flanges	Diam. Of Bolt Circle	No. Of Bolts	Diam. Of Bolts	Length Of Bolts
1	4⅞	3½	4	⅝	2¼
1¼	5¼	3⅞	4	⅝	2½
1½	6⅛	4½	4	¾	2½
2	6½	5	8	⅝	2½
2½	7½	5⅞	8	¾	3
3	8¼	6⅝	8	¾	3¼
3½	9	7¼	8	¾	3¼
4	10	7⅞	8	¾	3½
5	11	9¼	8	¾	3¾
6	12½	10⅝	12	¾	3¾
8	15	13	12	⅞	4¼
10	17½	15¼	16	1	5
12	20½	17¾	16	1⅛	5½
14 O.D.	23	20¼	20	1⅛	5¾
16 O.D.	25½	22½	20	1¼	6

FEET HEAD OF WATER TO PSI

Feet Head	Pounds Per Square Inch	Feet Head	Pounds Per Square Inch
1	.43	100	43.31
2	.87	110	47.64
3	1.30	120	51.97
4	1.73	130	56.30
5	2.17	140	60.63
6	2.60	150	64.96
7	3.03	160	69.29
8	3.46	170	73.63
9	3.90	180	77.96
10	4.33	200	86.62
15	6.50	250	108.27
20	8.66	300	129.93
25	10.83	350	151.58
30	12.99	400	173.24
40	17.32	500	216.55
50	21.65	600	259.85

FEET HEAD OF WATER TO PSI (Cont'd)

Feet Head	Pounds Per Square Inch	Feet Head	Pounds Per Square Inch
60	25.99	700	303.16
70	30.32	800	346.47
80	34.65	900	389.78
90	38.98	1000	433.00

NOTE: One foot of water at 62° Fahrenheit equals .433 pound pressure per square inch. To find the pressure per square inch for any feet head not given in the table above, multiply the feet head by .433.

WATER PRESSURE TO FEET HEAD

Pounds Per Square Inch	Feet Head	Pounds Per Square Inch	Feet Head
1	2.31	100	230.90
2	4.62	110	253.98
3	6.93	120	277.07
4	9.24	130	300.16
5	11.54	140	323.25
6	13.85	150	346.34
7	16.16	160	369.43
8	18.47	170	392.52
9	20.78	180	415.61
10	23.09	200	461.78
15	34.63	250	577.24
20	46.18	300	692.69
25	57.72	350	808.13
30	69.27	400	922.58
40	92.36	500	1154.48
50	115.45	600	1385.39
60	138.54	700	1616.30
70	161.63	800	1847.20
80	184.72	900	2078.10
90	207.81	1000	2309.00

NOTE: One pound of pressure per square inch of water equals 2.309 feet of water at 62° Fahrenheit. Therefore, to find the feet head of water for any pressure not given in the table above, multiply the pressure pounds per square inch by 2.309.

BOILING POINTS OF WATER
AT VARIOUS PRESSURES

Vacuum, In Inches Of Mercury	Boiling Point	Vacuum, In Inches Of Mercury	Boiling Point
29	76.62	7	198.87
28	99.93	6	200.96
27	114.22	5	202.25
26	124.77	4	204.85
25	133.22	3	206.70
24	140.31	2	208.50
23	146.45	1	210.25
22	151.87	Gauge Lbs.	
21	156.75	0	212.
20	161.19	1	215.6
19	165.24	2	218.5
18	169.00	4	224.4
17	172.51	6	229.8
16	175.80	8	234.8
15	178.91	10	239.4
14	181.82	15	249.8
13	184.61	25	266.8
12	187.21	50	297.1
11	189.75	75	320.1
10	192.19	100	337.9
9	194.50	125	352.9
8	196.73	200	387.9

TOTAL THERMAL EXPANSION OF
PIPING MATERIAL IN INCHES
PER 100 FT. ABOVE 32°F.

Temperature °F	Carbon And Carbon Moly Steel	Cast Iron	Copper	Brass And Bronze	Wrought Iron
32	0	0	0	0	0
100	0.5	0.5	0.8	0.8	0.5
150	0.8	0.8	1.4	1.4	0.9
200	1.2	1.2	2.0	2.0	1.3
250	1.7	1.5	2.7	2.6	1.7
300	2.0	1.9	3.3	3.2	2.2
350	2.5	2.3	4.0	3.9	2.6
400	2.9	2.7	4.7	4.6	3.1
450	3.4	3.1	5.3	5.2	3.6
500	3.8	3.5	6.0	5.9	4.1
550	4.3	3.9	6.7	6.5	4.6

TOTAL THERMAL EXPANSION OF PIPING MATERIAL IN INCHES PER 100 FT. ABOVE 32°F. (Cont'd)

Temper- ature °F	Carbon And Carbon Moly Steel	Cast Iron	Copper	Brass And Bronze	Wrought Iron
600	4.8	4.4	7.4	7.2	5.2
650	5.3	4.8	8.2	7.9	5.6
700	5.9	5.3	9.0	8.5	6.1
750	6.4	5.8	—	—	6.7
800	7.0	6.3	—	—	7.2
850	7.4	—	—	—	—
900	8.0	—	—	—	—
950	8.5	—	—	—	—
1000	9.	—	—	—	—

SPECIFIC GRAVITY OF GASES
(At 60°F and 29.92″ Hg)

Dry air (1 cu. ft. at 60°F. and 29.92″ Hg. weighs .07638 pound) 1.000

Acetylene	C_2H_2	0.91
Ethane	C_2H_6	1.05
Methane	CH_4	0.554
Ammonia	NH_3	0.596
Carbon-dioxide	CO_2	1.53
Carbon-monoxide	CO	0.967
Butane	C_4H_{10}	2.067
Butene	C_4H_8	1.93
Chlorine	Cl_2	2.486
Helium	He	0.138
Hydrogen	H_2	0.0696
Nitrogen	N_2	0.9718
Oxygen	O_2	1.1053

SPECIFIC GRAVITY OF LIQUIDS

Liquid	Temp. °F	Specific Gravity
Water (1 cu.-ft. weighs 62.41 lb.)	50	1.00
Brine (Sodium Chloride 25%)	32	1.20
Pennsylvania Crude Oil	80	0.85
Fuel Oil No. 1 and 2	85	0.95
Gasoline	80	0.74
Kerosene	85	0.82
Lubricating Oil SAE 10-20-30	115	0.94

TYPICAL BTU VALUES OF FUELS

ASTM Rank Solids	BTU Values Per Pound
Anthracite Class I	11,230
Bituminous Class II Group 1	14,100
Bituminous Class II Group 3	13,080
Sub-Bituminous Class III Group 1	10,810
Sub-Bituminous Class III Group 2	9,670

Liquids	BTU Values Per Gal.
Fuel Oil No. 1	138.870
Fuel Oil No. 2	143,390
Fuel Oil No. 4	144,130
Fuel Oil No. 5	142,720
Fuel Oil No. 6	137,275

Gases	BTU Values Per Cu. Ft.
Natural Gas	935 to 1132
Producers Gas	163
Illuminating Gas	534
Mixed (Coke oven and water gas)	545

STANDARD TABLES OF METRIC MEASURE

Linear Measure

Unit	Value in meters	Symbol or Abbrev.
Micron	0.000001	μ
Millimeter	0.001	mm.
Centimeter	0.01	cm.
Decimeter	0.1	dm.
Meter (unit)	1.0	m.
Dekameter	10.0	dkm.
Hectometer	100.0	hm.
Kilometer	1,000.00	km.
Myriameter	10,000.0	Mm.
Megameter	1,000,000.0	

Volume

Unit	Value in liters	Symbol or Abbrev.
Milliliter	0.001	ml.
Centiliter	0.01	cl.
Deciliter	0.1	dl.
Liter (unit)	1.0	l.
Dekaliter	10.0	dkl.
Hectoliter	100.0	hl.
Kiloliter	1,000.0	kl.

Surface Measure

Unit	Value in square meters	Symbol or Abbrev.
Square milli- meter	0.000001	mm.^{2}
Square centi- meter	0.0001	cm.^{2}
Square deci- meter	0.01	dm.^{2}
Square meter (centiare)	1.0	m.^{2}
Square dekameter (are)	100.0	a.^{2}
Hectare	10,000.0	ha.^{2}
Square kilometer	1,000,000.0	km^2

Mass

Unit	Value in grams	Symbol or Abbrev.
Microgram	0.000001	μg.
Milligram	0.001	mg.
Centigram	0.01	cg.
Decigram	0.1	dg.
Gram (unit)	1.0	g.
Dekagram	10.0	dkg.
Hectogram	100.0	hg.
Kilogram	1,000.0	kg.
Myriagram	10,000.0	Mg.
Quintal	100,000.0	q.
Ton	1,000,000.0	

Cubic Measure

Unit	Value in cubic meters	Symbol or Abbrev.
Cubic micron	10^{-18}	μ^3
Cubic millimeter	10^{-9}	mm.³
Cubic centimeter	10^{-6}	cm.³
Cubic decimeter	10^{-3}	dm.³
Cubic meter	1	m.³
Cubic dekameter	10^3	dkm.³
Cubic hectometer	10^6	hm.³
Cubic kilometer	10^9	km.³

Decimal and Millimeter Equivalents

Millimeter Equivalents of Decimals (0.01″ to 0.99″)

Dec.	0	1	2	3	4	5	6	7	8	9
0.0	....	0.254	0.508	0.762	1.016	1.270	1.524	1.778	2.032	2.286
0.1	2.540	2.794	3.048	3.302	3.556	3.810	4.064	4.318	4.572	4.826
0.2	5.080	5.334	5.588	5.842	6.096	6.350	6.604	6.858	7.112	7.366
0.3	7.620	7.874	8.128	8.392	8.636	8.890	9.144	9.398	9.652	9.906
0.4	10.160	10.414	10.688	10.922	11.176	11.430	11.684	11.938	12.192	12.446
0.5	12.700	12.954	13.208	13.462	13.716	13.970	14.224	14.478	14.732	14.986
0.6	15.240	15.494	15.748	16.022	16.256	16.510	16.764	17.018	17.272	17.526
0.7	17.780	18.034	18.288	18.542	18.796	19.050	19.304	19.558	19.812	20.066
0.8	20.320	20.574	20.828	21.082	21.336	21.590	21.844	22.098	22.352	22.606
0.9	22.860	23.114	23.368	23.622	23.876	24.130	24.384	24.638	24.892	25.146

Example 0.1″ = 2.540 mm., 0.75″ = 19.050 mm.

Decimal Equivalents of Millimeters (1 mm. to 99 mm.)

Mm.	0	1	2	3	4	5	6	7	8	9
0	....	0.0394	0.0787	0.1181	0.1575	0.1968	0.2362	0.2756	0.3150	0.3543
1	0.3937	0.4331	0.4724	0.5118	0.5512	0.5906	0.6299	0.6693	0.7087	0.7480
2	0.7874	0.8268	0.8661	0.9055	0.9449	0.9842	1.0236	1.0630	1.1024	1.1417
3	1.1811	1.2205	1.2598	1.2992	1.3386	1.3780	1.4173	1.4567	1.4961	1.5354
4	1.5748	1.6142	1.6535	1.6929	1.7323	1.7716	1.8110	1.8504	1.8898	1.9291
5	1.9685	2.0079	2.0472	2.0866	2.1260	2.1654	2.2047	2.2441	2.2835	2.3228
6	2.3622	2.4016	2.4409	2.4803	2.5197	2.5590	2.5984	2.6378	2.6772	2.7165
7	2.7559	2.7953	2.8346	2.8740	2.9134	2.9528	2.9921	3.0315	3.0709	3.1102
8	3.1496	3.1890	3.2283	3.2677	3.3071	3.3464	3.3858	3.4252	3.4646	3.5039
9	3.5433	3.5827	3.6220	3.6614	3.7008	3.7402	3.7795	3.8189	3.8583	3.8976

Example 10 mm. = 0.3937″, 57 mm. = 2.2441″

Conversion of English and Metric Measures

Linear Measure

Unit	Inches to millimeters	Millimeters to inches	Feet to meters	Meters to feet	Yards to meters	Meters to yards	Miles to kilometers	Kilometers to miles
1	25.40	0.03937	0.3048	3.281	0.9144	1.094	1.609	0.6214
2	50.80	0.07874	0.6096	6.562	1.829	2.187	3.219	1.243
3	76.20	0.1181	0.9144	9.842	2.743	3.281	4.820	1.864
4	101.60	0.1575	1.219	13.12	3.658	4.374	6.437	2.485
5	127.00	0.1968	1.524	16.40	4.572	5.468	8.047	3.107
6	152.40	0.2362	1.829	19.68	5.486	6.562	9.656	3.728
7	177.80	0.2756	2.134	22.97	6.401	7.655	11.27	4.350
8	203.20	0.3150	2.438	26.25	7.315	8.749	12.87	4.971
9	228.60	0.3543	2.743	29.53	8.230	9.842	14.48	5.592

Example   1 in. = 2540 mm., 1 m. = 3.281 ft., 1 Km. = 0.6214 mi.

Surface Measure

Unit	Square inches to square centimeters	Square centimeters to square inches	Square feet to square meters	Square meters to square feet	Square yards to square meters	Square meters to square yards	Acres to hectares	Hectares to acres	Square miles to square kilometers	Square kilometers to square miles
1	6.452	0.1550	0.0929	10.76	0.8361	1.196	0.4047	2.471	2.59	0.3861
2	12.90	0.31	0.1859	21.53	1.672	2.392	0.8094	4.942	5.18	0.7722
3	19.356	0.465	0.2787	32.29	2.508	3.588	1.214	7.413	7.77	1.158
4	25.81	0.62	0.3716	43.06	3.345	4.784	1.619	9.884	10.36	1.544
5	32.26	0.775	0.4645	53.82	4.181	5.98	2.023	12.355	12.95	1.931
6	38.71	0.93	0.5574	64.58	5.017	7.176	2.428	14.826	15.54	2.317
7	45.16	1.085	0.6503	75.35	5.853	8.372	2.833	17.297	18.13	2.703
8	51.61	1.24	0.7432	86.11	6.689	9.568	3.237	19.768	20.72	3.089
9	58.08	1.395	0.8361	96.87	7.525	10.764	3.642	22.239	23.31	3.475

Example 1 sq. in. = 6.452 sq. cm., 1 sq. m. = 1.196 sq. yds.,
1 sq. mi. = 2.59 sq. Km.

Cubic Measure

Unit	Cubic inches to cubic centimeters	Cubic centimeters to cubic inches	Cubic feet to cubic meters	Cubic meters to cubic feet	Cubic yards to cubic meters	Cubic meters to cubic yards	Gallons to cubic feet	Cubic feet to gallons
1	16.39	0.06102	0.02832	35.31	0.7646	1.308	0.1337	7.481
2	32.77	0.1220	0.05663	70.63	1.529	2.616	0.2674	14.96
3	49.16	0.1831	0.08495	105.9	2.294	3.924	0.4010	22.44
4	65.55	0.2441	0.1133	141.3	3.058	5.232	0.5347	29.92
5	81.94	0.3051	0.1416	176.6	3.823	6.540	0.6684	37.40

Cubic Measure (Cont'd)

Unit	Cubic inches to cubic centimeters	Cubic centimeters to cubic inches	Cubic feet to cubic meters	Cubic meters to cubic feet	Cubic yards to cubic meters	Cubic meters to cubic yards	Gallons to cubic feet	Cubic feet to gallons
6	98.32	0.3661	0.1699	211.9	4.587	7.848	0.8021	44.88
7	114.7	0.4272	0.1982	247.2	5.352	9.156	0.9358	52.36
8	131.1	0.4882	0.2265	282.5	6.116	10.46	1.069	59.84
9	147.5	0.5492	0.2549	371.8	6.881	11.77	1.203	67.32
Example 1 cu. cm. = 0.06102 cu. in., 1 gal. = 0.1337 cu. ft.								

Volume or Capacity Measure

Unit	Liquid ounces to cubic centimeters	Cubic centimeters to liquid ounces	Pints to liters	Liters to pints	Quarts to liters	Liters to quarts	Gallons to liters	Liters to gallons	Bushels to hectoliters	Hectoliters to bushels
1	29.57	0.03381	0.4732	2.113	0.9463	1.057	3.785	0.2642	0.3524	2.838
2	59.15	0.06763	0.9463	4.227	1.893	2.113	7.571	0.5284	0.7048	5.676
3	88.72	0.1014	1.420	6.340	2.839	3.785	11.36	0.7925	1.057	8.513
4	118.3	0.1353	1.893	8.454	3.170	4.227	15.14	1.057	1.410	11.35
5	147.9	0.1691	2.366	10.57	4.732	5.284	18.93	1.321	1.762	14.19
6	177.4	0.2029	2.839	12.68	5.678	6.340	22.71	1.585	2.114	17.03
7	207.0	0.2367	3.312	14.79	6.624	7.397	26.50	1.849	2.467	19.86
8	236.6	0.2705	3.785	16.91	7.571	8.454	30.28	2.113	2.819	22.70
9	266.2	0.3043	4.259	19.02	8.517	9.510	34.07	2.378	3.171	25.54
Example 1 l. = 2.113 pts., 1 gal = 3.785 l.										

METRIC BALL BEARING DIMENSIONS
EXTRA LIGHT "100" SERIES

Basic Bearing Number	Bore		O.D.		Width	
	mm	Inches	mm	Inches	mm	Inches
100	10	0.3937	26	1.0236	8	0.3150
101	12	0.4724	28	1.1024	8	0.3150
102	15	0.5906	32	1.2598	9	0.3543
103	17	0.6693	35	1.3780	10	0.3937
104	20	0.7874	42	1.6535	12	0.4724
105	25	0.9843	47	1.8504	12	0.4724
106	30	1.1811	55	2.1654	13	0.5118
107	35	1.3780	62	2.4409	14	0.5512
108	40	1.5748	68	2.6772	15	0.5906
109	45	1.7717	75	2.9528	16	0.6299
110	50	1.9685	80	3.1496	16	0.6299
111	55	2.1654	90	3.5433	18	0.7087

METRIC BALL BEARING DIMENSIONS
EXTRA LIGHT "100" SERIES (Cont'd)

Basic Bearing Number	Bore		O.D.		Width	
	mm	Inches	mm	Inches	mm	Inches
112	60	2.3622	95	3.7402	18	0.7087
113	65	2.5591	100	3.9370	18	0.7087
114	70	2.7559	110	4.3307	20	0.7874
115	75	2.9528	115	4.5276	20	0.7874
116	80	3.1496	125	4.9213	22	0.8661
117	85	3.3465	130	5.1181	22	0.8661
118	90	3.5433	140	5.5118	24	0.9449
119	95	3.7402	145	5.7087	24	0.9449
120	100	3.9370	150	5.9055	24	0.9449
121	105	4.1339	160	6.2992	26	1.0236

METRIC BALL BEARING DIMENSIONS
LIGHT "200" SERIES

Basic Bearing Number	Bore		O.D.		Width	
	mm	Inches	mm	Inches	mm	Inches
200	10	0.3937	30	1.1811	9	0.3543
201	12	0.4724	32	1.2598	10	0.3937
202	15	0.5906	35	1.3780	11	0.4331
203	17	0.6693	40	1.5748	12	0.4724
204	20	0.7874	47	1.8504	14	0.5512
205	25	0.9843	52	2.0472	15	0.5906
206	30	1.1811	62	2.4409	16	0.6299
207	35	1.3780	72	2.8346	17	0.6653
208	40	1.5748	80	3.1496	18	0.7087
209	45	1.7717	85	3.3465	19	0.7480
210	50	1.9685	90	3.5433	20	0.7874
211	55	2.1654	100	3.9370	21	0.8268
212	60	2.3633	110	4.3307	22	0.8661
213	65	2.5591	120	4.7244	23	0.9055
214	70	2.7559	125	4.9213	24	0.9449
215	75	2.9528	130	5.1181	25	0.9843
216	80	3.1496	140	5.5118	26	1.0236
217	85	3.3465	150	5.9055	28	1.1024
218	90	3.5433	160	6.2992	30	1.1811
219	95	3.7402	170	6.6929	32	1.2598
220	100	3.9370	180	7.0866	34	1.3386
221	105	4.1339	190	7.4803	36	1.4137
222	110	4.3307	200	7.8740	38	1.4961

METRIC BALL BEARING DIMENSIONS
MEDIUM "300" SERIES

Basic Bearing Number	Bore		O.D.		Width	
	mm	Inches	mm	Inches	mm	Inches
300	10	0.0397	35	1.3780	11	0.4331
301	12	0.4724	37	1.4567	12	0.4724
302	15	0.5906	42	1.6535	13	0.5118
303	17	0.6693	47	1.8504	14	0.5512
304	20	0.7874	52	2.0472	15	0.5906
305	25	0.9843	62	2.4409	17	0.6693
306	30	1.1811	72	2.8346	19	0.7480
307	35	1.3780	80	3.1496	21	0.8268
308	40	1.5748	90	3.5433	23	0.9055
309	45	1.7717	100	3.9370	25	0.9843
310	50	1.9685	110	4.3307	27	1.0630
311	55	2.1654	120	4.7244	29	1.1417
312	60	2.3622	130	5.1181	31	1.2205
313	65	2.5591	140	5.5118	33	1.2992
314	70	2.7559	150	5.9055	35	1.3780
315	75	2.9528	160	6.2992	37	1.4567
316	80	3.1469	170	6.6929	39	1.5354
317	85	3.3465	180	7.0866	41	1.6142
318	90	3.5433	190	7.4803	43	1.6929
319	95	3.7402	200	7.8740	45	1.7717
320	100	3.9370	215	8.4646	47	1.8504
321	105	4.1339	225	8.8583	49	1.9291
322	110	4.3307	240	9.4480	50	1.9685
324	120	4.7244	260	10.2362	55	2.1654
326	130	5.1181	280	11.0236	58	2.2835
328	140	5.5118	300	11.8110	62	2.4409
330	150	5.9055	320	12.5984	65	2.5591
332	160	6.2992	340	13.3858	68	2.6772
334	170	6.6929	360	14.1732	72	2.8346
336	180	7.0866	380	14.9606	75	2.9528
338	190	7.4803	400	15.7480	78	3.0709
340	200	7.8740	420	16.5354	80	3.1496
342	210	8.2677	440	17.3228	84	3.3071
344	220	8.6614	460	18.1002	88	3.4646
348	240	9.4488	500	19.6850	95	3.7402
352	260	10.2362	540	21.2598	102	4.0157
356	280	11.0236	580	22.8346	108	4.2520

METRIC BALL BEARING DIMENSIONS
HEAVY "400" SERIES

Basic Bearing Number	Bore		O.D.		Width	
	mm	Inches	mm	Inches	mm	Inches
403	17	0.6693	62	2.4409	17	0.6693
404	20	0.7874	72	2.8345	19	0.7480
405	25	0.9843	80	3.1496	21	0.8268
406	30	1.1811	90	3.5433	23	0.9055
407	35	1.3780	100	3.9370	25	0.9843
408	40	1.5748	110	4.3307	27	1.0630
409	45	1.7717	120	4.7244	29	1.1417
410	50	1.9685	130	5.1181	31	1.2205
411	55	2.1654	140	5.5118	33	1.2992
412	60	2.3622	150	5.9055	35	1.3780
413	65	2.5591	160	6.2992	37	1.4567
414	70	2.7559	180	7.0866	42	1.6535
415	75	2.9528	190	7.4803	45	1.7717
416	80	3.1496	200	7.8740	48	1.8898
417	85	3.3465	210	8.2677	52	2.0472
418	90	3.5433	225	8.8533	54	2.1260
419	95	3.7402	250	9.8425	55	2.1654
420	100	3.9370	265	10.4331	60	2.3622
421	105	4.1339	290	11.4173	65	2.5591
422	110	4.3307	320	12.5984	70	2.5759

Tap Drill Sizes For American Standard Threads

Diam. of Thread	Threads per Inch	Drill*	Decimal Equiv.
No. 0—.060	80 NF	3/64	.0469
1—.073	64 NC	1.5 MM	.0591
	72 NF	53	.0595
2—.086	56 NC	50	.0700
	64 NF	50	.0700
3—.099	48 NC	5/64	.0781
	56 NF	45	.0820
4—.112	40 NC	43	.0890
	48 NF	42	.0935
5—.125	40 NC	38	.1015
	44 NF	37	.1040

Tap Drill Sizes For American Standard Threads (Cont'd)

Diam. of Thread	Threads per inch	Drill*	Decimal Equiv.
6—.138	32 NC	36	.1065
	40 NF	33	.1130
8—.164	32 NC	29	.1360
	36 NF	29	.1360
10—.190	24 NC	25	.1495
	32 NF	21	.1590
12—.216	24 NC	16	.1770
	28 NF	14	.1820
¼	20 NC	7	.2010
	28 NF	3	.2130
	32 NEF	$\frac{7}{32}$	.2188
$\frac{5}{16}$	18 NC	F	.2570
	24 NF	I	.2720
	32 NEF	$\frac{9}{32}$	.2812
⅜	16 NC	$\frac{5}{16}$	.3125
	24 NF	Q	.3320
	32 NEF	$\frac{11}{32}$	.3438
$\frac{7}{16}$	14 NC	U	.3680
	20 NF	$\frac{25}{64}$	.3906
	28 NEF	Y	.4040
½	12 N	$\frac{27}{64}$	.4219
	13 NC	$\frac{27}{64}$	.4219
	20 NF	$\frac{29}{64}$	.4531
	28 NEF	$\frac{15}{32}$	.4687
$\frac{9}{16}$	12 NC	$\frac{31}{64}$	.4844
	18 NF	$\frac{33}{64}$	.5156
	24 NEF	$\frac{33}{64}$	.5156
⅝	11 NC	$\frac{17}{32}$	.5312
	12 N	$\frac{35}{64}$	.5469
	18 NF	14.5 MM	.5709
	24 NEF	$\frac{37}{64}$	.5781
$\frac{11}{16}$	12 N	$\frac{39}{64}$	.6094
	24 NEF	16.5 MM	.6496
	10 NC	16.5 MM	.6496
¾	12 N	17 MM	.6693
	16 NF	17.5 MM	.6890
	20 NEF	$\frac{45}{64}$	.7031

Tap Drill Sizes For American Standard Threads (Cont'd)

Diam. of Thread	Threads per Inch	Drill*	Decimal Equiv.
13/16	12 N	18.5 MM	.7283
	16 N	3/4	.7500
	20 NEF	49/64	.7656
7/8	9 NC	49/64	.7656
	12 N	20 MM	.7874
	14 NF	25.5 MM	.8071
	16 N	13/16	.8125
	20 NEF	21 MM	.8268
15/16	12 N	53/64	.8594
	16 N	7/8	.8750
	20 NEF	22.5 MM	.8858
1	8 NC	7/8	.8750
	12 N	59/64	.9219
	14 NF	23.5 MM	.9252
	16 N	15/16	.9375
	20 NEF	61/64	.9531
1½	6 NC	1 21/64	1.3281
	8 N	1 3/8	1.3750
	12 NF	36 MM	1.4173
	16 N	1 7/16	1.4375
	18 NEF	1 29/64	1.4531
2	4½ NC	1 25/32	1.7812
	8 N	1 7/8	1.8750
	12 N	1 59/64	1.9219
	16 NEF	1 15/16	1.9375
2½	4 NC	2 1/4	2.2500
	8 N	2 3/8	2.3750
	12 N	61.5 MM	2.4213
	16 N	2 7/16	2.4375
3	4 NC	2 3/4	2.7500
	8 N	2 7/8	2.8750
	12 N	74 MM	2.9134
	16 N	2 15/16	2.9375

*To produce approximately 75% full thread

TEMPERING AND HEAT COLORS

	Color	Degrees	
		Fahrenheit	Centigrade
Temper Colors	Faint straw	400	205
	Straw	440	225
	Deep straw	475	245
	Bronze	520	270
	Peacock blue	540	280
	Full blue	590	310
	Light blue	640	340
Heat Colors	Faint red	930	500
	Blood red	1075	580
	Dark cherry	1175	635
	Medium cherry	1275	690
	Cherry	1375	745
	Bright cherry	1450	790
	Salmon	1550	840
	Dark orange	1680	890
	Orange	1725	940
	Lemon	1830	1000
	Light yellow	1975	1080
	White	2200	1200